# Nuclear Reactions in Stellar Surfaces and their Relations with Stellar Evolution

**Topics in Astrophysics and Space Physics**

Edited by A. G.W. Cameron, *Yeshiva University*, and
G. B. Field, *University of California at Berkeley*

H. REEVES   *Nuclear Reactions in Stellar Surfaces and
their Relations with Stellar Evolution*

Additional volumes in the series:

T. ARNY   *Star Formation in Interstellar Clouds*
C. R. COWLEY   *The Theory of Stellar Spectra*
V. L. GINZBURG   *Elementary Processes for Cosmic Ray Astrophysics*
V. L. GINZBURG   *The Origin of Cosmic Rays*
S. GLASSTONE   *The Book of Mars*
D. B. MELROSE   *Plasma Astrophysics*
K. GREISEN   *The Physics of X-ray, Gamma-ray and Particle Sources*

# Nuclear Reactions in Stellar Surfaces and their Relations with Stellar Evolution

HUBERT REEVES

*Institut d'Astrophysique de Paris
and Centre d'Etudes Nucléaires de Saclay*

GORDON AND BREACH SCIENCE PUBLISHERS

NEW YORK          LONDON          PARIS

# Preface

THIS BOOK IS based on a series of lectures given at the Winter School of the University of Tel Aviv, Israel (Jan. 1969) and at the "Cours de Structure Interne" de l'Association Vaudoise des Chercheurs en Physique, at Saas Fee, Switzerland (March 1969).

Much of the material presented here comes from a fruitful collaboration of several years with scientists at the Institut de Physique Nucléaire, Orsay; at the Institut d'Astrophysique, Paris and at the Service d'Electronique Physique, Saclay. I want to mention in particular:

Jean AUDOUZE
René BERNAS
Marcelle EPHERRE
Elie GRADSZTAJN
Charles RYTER
Evry SCHATZMAN
Françoise YIOU

June 1970                                    HUBERT REEVES

v

# Preliminary

The one nice thing about writing your lecture notes after you have given the courses (and not before as usually required but seldom done) is that the reactions and remarks of the students have plenty of time to sink in. Only after the final survey of these notes, did I become fully convinced that the agent responsible for the production of the atoms of lithium beryllium and boron in stars and in the solar system was the galactic cosmic ray flux bombarding the instellar gas throughout the life of the galaxy (as discussed in chapter V of these notes).

Shortly thereafter, in July 1969, I visited the Institute of Theoretical Astronomy in Cambridge where I expected to find the proponents of the stellar-flare origin of these elements. However, the situation turned out to be different, as I found that a similar change of outlok had already taken place in a number of people's mind. As a result Dr. Fred Hoyle, W. A. Fowler and myself have since presented our new theory in *Nature* **226,** 767 (1970), which could serve as a complement to these lecture notes.

# Contents

Preface    .    .    .    .    .    .    .    .    .    v

Preliminary .    .    .    .    .    .    .    .    .    vii

**CHAPTER I  Introduction**    .    .    .    .    .    .    .    1

I-1   Purpose of the book  1

I-2   A bird's eye view of the subject; the nuclear properties
of the L-elements and their nucleosynthesis  1

**CHAPTER II  Nuclear Phenomena**    .    .    .    .    .    .    4

II-1  Nuclear reactions and their relations with nuclear properties; a brief phenomenological analysis  4

      II-1-1  The mass defect  4

      II-1-2  The two-step model of nuclear reactions  7

      II-1-3  Experimental results: the mass yield  13

      II-1-4  Experimental results: the isobaric yield  15

      II-1-5  The experimental results as seen from the point
of view of the statistical theory  16

II-2  High-energy nuclear reactions in stars  17

      II-2-1  The cross-sections for formation of the light
elements; the formation ratios in stellar
flares  17

      II-2-1a  The L-elements  17

      II-2-1b  The isotopes He, $d$, $n$  23

      II-2-2  The destruction cross-sections of light
elements  25

      II-2-3  Where do the destruction reactions take
place?  29

      II-2-4  The time variation of the L-element
abundances  32

CHAPTER III   Atomic Phenomena: The Energetics of L-element For-
              mation    .    .    .    .    .    .    .    .    .    34

      III-1   Energy conservation and energy transformation   34

      III-2   The "stopping power" of matter and the range of fast
              protons in hydrogen   35
              III-2-1  The case of a neutral gas   35
              III-2-2  The case of an ionized gas (plasma)   37
      III-3   The computation of the energy requirement for L-
              element formation   39


CHAPTER IV    Stellar Observations and Interpretations   .    .    .    41

      IV-1    Stellar observations of lithium   41
      IV-2    Interpretation of the lithium data   42
      IV-3    Correlation with other stellar properties   51
      IV-4    The stellar dynamo   54
      IV-5    Stellar observations of the $^7Li/^6Li$ ratio   56
      IV-6    Stellar observations of beryllium   57
      IV-7    An interpretation of the combined Li–Be data   59
      IV-8    The D/H and $^3He/^4He$ ratio in stars   61
      IV-9    The Sun and the Solar system   62
      IV-10   The Stellar energy problem and its implications   67


CHAPTER V     Theories of the Origin of the Light elements   .    .    .    71

      V-1     The thermonuclear theories   72
      V-2     The spallative galactogenic theories   74
              V-2-1    The effect of the galactic cosmic-ray
                       protons   75
              V-2-2    The deceleration of the L-elements   76
      V-3     The spallative autogenic theory   76


CHAPTER VI    Summary   .    .    .    .    .    .    .    .    .    78


              Appendix   .    .    .    .    .    .    .    .    .    80

              Cross section for the reaction $p + d \rightarrow p + p + n$   80
              Main sequence stars   82


              References   .    .    .    .    .    .    .    .    .    83

# CHAPTER I

# Introduction

## I-1  PURPOSE OF THE BOOK

IT IS GENERALLY believed today that high-energy nuclear reactions induced by fast protons are responsible for the generation of most of the L-elements (Li, Be, B) present in stellar photosphere (spallative origin).

It is also believed that these nuclear reactions take place in the very star in which they are observed (autogenic origin).

In the first four sections of this course these views will be adopted. In section V the various theories of L-element formation will be reviewed and discussed.

Our task in this book will be to compare the laboratory data (nuclear data on the formation and destruction processes; atomic data on energy losses) with the astronomical data (stellar abundances and their correlations with a number of astrophysical parameters).

Such a comparison should reveal some features of surface stellar processes, should throw some light on the relation between surfaces and interiors, and, by so doing, lead to further understanding of stellar evolution.

Much remains to be done on this "program". Nevertheless it is already of great interest to look at the situation as it stands now.

## I-2  A BIRD'S EYE VIEW OF THE SUBJECT:
## THE NUCLEAR PROPERTIES OF THE L-ELEMENTS
## AND THEIR NUCLEOSYNTHESIS

Many features of the universal abundance curve of the elements can be qualitatively understood through a knowledge of the nuclear properties of these elements; the major (iron) peak corresponding to the most stable elements; the secondary peaks corresponding to the nuclei with "magic" number of neutrons, and, also to the light nuclei with an integer number of alpha-particles, etc... In this respect it is fruitful to begin our study with a brief review of the nuclear properties of the L-elements ($4 < A < 12$).

The most important factor here is the large binding energy (and stability) of alpha-clusters (2 protons and 2 neutrons) as compared to any other nuclear arrangement. As a result every nucleus in this mass range has a rather

1

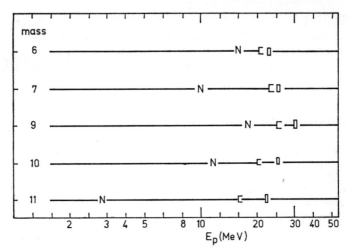

**Figure I-1**   Minimum threshold energies ($Q$ values) for the formation of
the L-elements by spallation reactions of fast protons on the targets $^{12}C$,
$^{14}N$, $^{16}O$ (identified as C, N, O). Note the large (endothermic) values
corresponding to the formation of mass-9 ($^9Be$)

precarious stability toward a rearrangement involving $He^4$ nuclei. For
instance no mass-5 nucleus (with five nucleons) manage to be particle-stable;
the lifetime is $\simeq 10^{-21}$ sec. The isotopes $^8Li$ and $^8B$ are beta-unstable with
respect to $^8Be$ which quickly ($\simeq 10^{-16}$ sec) breaks into two alphas; hence
stable masse-8 nuclei are also nonexistent.

The situation in the mass-9 is slightly better, but yet remains deeply
marked by the alpha-stability; $^9B$ is unstable against $2\alpha + p$ ($\tau = 10^{-19}$ sec)
but $^9Be$ (barely) escapes disintegration (very small binding energy). This
weak stability is first reflected in the fact that the endothermic $Q^*$ values
corresponding to its formation (figure I-1) are remarkably large and the
(exothermic) $Q$ values corresponding to its destruction are remarkably small.

The weak stability of $^9Be$ is also reflected in the fact that it has only one
"bound" state: the ground state; all the excited states are unstable against
particle break-up.

The isotopes $^6Li$, $^7Li$ and $^{10}B$, $^{11}B$ do better, but remain all comparatively
fragile: in proton reactions they all quickly break in residues involving alpha-
particles.

---

* The $Q$ value is the energy needed (endothermic case) or released (exothermic case)
in a reaction.

Another important factor is the Coulomb repulsion associated with the nuclear charge. This factor is dominant in the very low energy range ($E \ll Z_1 Z_2 e^2 / R = B$) characteristic of thermonuclear reactions. At higher energies it becomes negligible.

More will be said about nuclear reactions in the next chapter. However from the preceding paragraphs the following discussion can already be made.

The isotopes $^6$Li, $^7$Li, $^9$Be, $^{10}$B, $^{11}$B are observed in various natural spots; their abundance ratios usually differ by one, or sometime by two orders of magnitude, not more.

In purely thermonuclear contexts ("big bang", "little bangs", stellar interiors) the difference in the (Coulomb dominated) formation and destruction reaction rates of the various L-elements are very large (see figure II-17 for instance). Correspondingly, relative abundances in very large ratios ($10^6$ or more) can reasonably be expected (and are actually found in appropriate calculation, for example in figure V-1).

At higher energies ($\simeq$ a few MeV) nuclear properties become important, and there, the similarity of the L-nuclei would considerably reduce the expected ratios. At still higher energies ($\simeq$ a few hundred MeV) the nucleus essentially becomes an aggregate of loose nucleons, forgetful of all individual properties, except, again, those related to the question of stability.

Hence, in a purely high-energy context rather small ratios of L-abundances are expected (one order of magnitude at most).

Comparison between these two extreme predictions and the natural abundances suggests that the dominant factor in their genesis should be the high-energy reactions, but also that the low-energy reactions should play a role. Indeed, after a more detailed analysis, we shall come to the conclusion that the stellar L-elements are formed by high-energy reactions and, later, partly destroyed by low-energy reactions.

As a confirmation to this view, in the galactic cosmic-rays, where *no* low-energy reactions (formation of destruction) are expected to take place, the L-element abundances differ by *less* than one order of magnitude (figure V-2).

This qualitative analysis is more than just a "hors d'œuvre" to what will come next. Stellar and meteoritic abundance ratios are still uncertain: in the last years both the stellar Be (Grevesse) and the meteoritic B determination (Quijano-Rico) have been altered by important factors. Much of what will follow may become useless if new "recalibrations" do take place, but the preceding qualitative analysis will presumably remain valid.

# CHAPTER II

# Nuclear Phenomena

## II-1  NUCLEAR REACTIONS AND THEIR RELATIONS WITH NUCLEAR PROPERTIES; A BRIEF PHENOMENOLOGICAL ANALYSIS

### II-1-1  The mass-defect

A nucleus is a cluster of $A$ nucleons (with $Z$ protons, $N$ neutrons, and a symmetry or "isospin" number $T_z = (N - Z)/2$, held together by nuclear forces.

Each nucleus has a mass $M$, and a mass-defect $\Delta M\,(Z, N)$

$$-\Delta M\,(Z, N) = [M - NM_n - ZM_p] \qquad \text{(II-0)}$$

which describes its stability with respect to the state in which the neutrons (of mass $M_n$) and the protons (of mass $M_p$) are dispersed. The term $\Delta M\,(N,Z)/A$ is called the "binding-energy per nucleon" of the nucleus.

Similarly, one can define a mass-defect and a binding-energy with respect to any other configuration. Of special interest to us are those involving alpha-particles.

The general properties of nuclear matter are best visualized through a three-dimensional diagram in which the function $\Delta M\,(N, Z)/A$ is plotted vertically above the two-dimensionnal $Z - N$ plane containing all the isotopes. At low $Z$ or $N$ value the valley of nuclear stability runs almost diagonally with rather steep slopes on both sides. At higher $Z$ or $N$ values the valley moves somewhat out of the diagonal (figure II-1); it widens and the slopes decrease progressively.

In figure II-2 the shape of the $(\Delta M/A)$ curve is shown along the valley, and in figure II-3. across the valley (at fixed $A$ and varying $T_Z = (N - Z)/2$.

In figure II-2 it appears that, in first approximation, the binding-energy of the nuclei is independent of $A$, $(\Delta M/A \simeq 8$ MeV in the valley). This is a reflection of the "short range" of the nuclear forces; a nucleus resembles more a drop of water (with short range Van der Waals forces) than an atom (with long-range electric forces).

A closer look at figure II-2 reveals immediately the deep minimum around Fe (responsible for the iron peak) and the overstability of $^4$He. The effect of

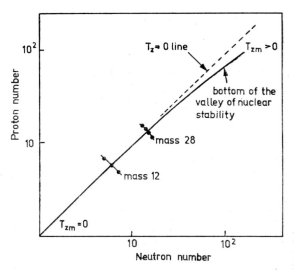

**Figure II-1**   Position of the valley of nuclear stability in the $Z - N$ plane. In figure II-2 the $\Delta M/A$ function is taken along the valley, and in figure II-3 along the isobaric lines at mass 12 and mass 28. Also shown is the line of neutron-proton symmetry $\{(T_Z = (N - Z)/2) = 0)\}$. The Coulomb repulsion is responsible for displacement of the valley on the neutron-excess side. $T_Z m$ is the value of $T_Z$ for the nuclei at the bottom of the valley

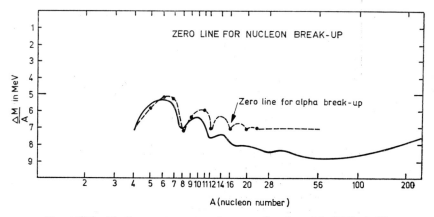

**Figure II-2**   Binding energy per nucleon as a function $A$ (solid line). The zero line for break-up in nucleons is on top. The dotted line is the zero line for break-up in the largest possible number of alphas. The binding energy per nucleon toward alpha break-up is the difference between the solid line and the dotted line. The weak stability against alpha break-up of the $L$ nuclei is clearly seen (the dotted line is then very close to the solid line)

1   Reeves (0296)

this overstability on the L-nuclei is best seen by a computation of the binding-energy per nucleon toward break-up in the largest possible number of alphas (for instance $^{11}B \to 2^4He + ^3He$, or $^{25}Mg \to 6^4He + n$, etc...) In figure II-2 this quantity is the difference between the solid line and the dotted line (which is the binding-energy of these alpha-containing aggregates for nucleon break-up; for instance $2^4He + ^3He \to 5p + 6n$, or $6^4He + n \to 12p + 13n$).

The unstability of masse five and eight (the dotted line is then *under* the solid line) and the weak stability of the other L-nuclei are well illustrated.

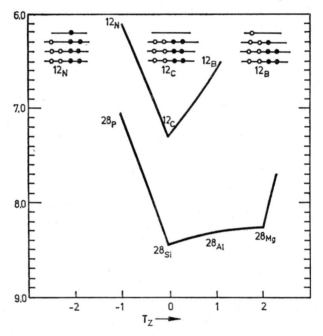

**Figure II-3**  Binding-energy per nucleon against nucleonic break-up for mass-12 and mass-28. The inserts show schematically the individual nucleon energy-levels (● protons, ⊙ neutrons). The effect of the generalized Pauli principle is well illustrated. The (weak) effect of the Coulomb energy appears through the lack of symmetry around $T_Z = 0$

On figure II-3 the binding-energy per nucleon is pictured for two masses: 12 and 28. The rise on both sides is very approximately proportional to $(T_Z - T_{Zm})^2$ where $T_{Zm}$ is the symmetry number of the most stable nucleus of given mass-number.

The rise is due to the effect of the "pairing energy": the binding energy is largest when the nucleons are "paired" four by four (neutron–proton; spin up–spin down). As we move away from symmetry, the generalized Pauli principle forbids certain pairing and sends some of the nucleons in higher energy levels (inserts in figure II-3).

The figure II-3 also shows the weak effect of the Coulomb energy: the curves are not completely symmetrical; the nuclei with larger charges have smaller binding energies. This effect is responsible for the displacement of the bottom of the valley at larger $A$ ($T_{Zm} > 0$) already visible for mass-28.

Many of the features of the mass-diagram will be of importance to us in our discussion of nuclear reaction rates.

## II-1-2   The two-step model of nuclear reactions

The processes taking place when an incident particle strikes a nucleus vary a great deal with the energy of this particle. However one feature remains in common all through the range of observed energies: the process takes place in two steps (we do not consider elastic* scattering in the present context).

In the first step, some energy is deposited in the structure, in a more or less efficient way, and with the possible loss of some of its nucleons.

In the second step the excited structure gets rid of its surplus energy, by particle, gamma and, occasionally, electron emission.

At low-energies the first step involves the capture of the incident particle. The excited structure thereby formed is called the "compound-nucleus" and the energy involves the masses and the incident kinetic energy. The de-excitation process is usually called "evaporation". Because of energy requirements, only a few channels are available: gamma-deexcitation, deexcitation by the entry-channel (compound-elastic scattering) and possibly few-particle channels.

The relative probability that the excited structure will deexcite by any one of these channels (or the "branching ratio") is generally a rapidly varying function of the excitation energy (resonant behaviour).

At higher energies (above 5 MeV or so) the resonances become very numerous; they start overlapping; and the probability of a given decay becomes a smooth function of the energy. Numerical values of the branching-ratios can be computed from nuclear statistical theories. The basic assump-

---

* The sophisticated reader will have surmized that I am talking about shape-elastic scattering. The compound-elastic scattering should be included in this discussion.

tion of these theories is the following: *the probability of a given channel is proportional to the phase space (kinetic, spin, isospin) occupied by all possible final states corresponding to this channel.*

The situation is reminiscent of the beta-decay theory: the probability of a certain group of states is the product of an "internal" factor times the number of states in this group. For weak interactions, the internal factor is (essentially) independent of energy (allowed transitions); hence the electron spectrum merely reflects the volume of phase space occupied by the outcoming particles.

In the case of nuclear interactions, the internal factors *may vary strongly with energy, but, when averaged over even narrow energy regions, they become rather weakly energy dependant,* as the studies of the so-called "strength functions" have shown repeatedly. In the frame of the statistical assumption, one simply neglects the energy dependance of nuclear effects (i.e. replaces them by geometrical and Coulomb factors). This way, the average value of nuclear reaction cross-sections can be estimated with reasonable accuracy (usually to a factor of two or less).

As we move up to higher energies (several tens of MeV) the nature of the nuclear processes changes somewhat; the nucleus becomes progressively

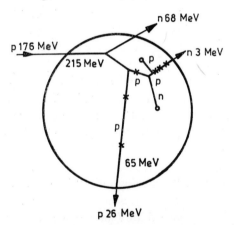

**Figure II-4**  A descriptive example of an intranuclear-cascade (from Friedlander, Kennedy, Miller). A proton of 176 MeV enters a nucleus and first sees its energy go up to 215 MeV by the collective effect of the nuclear forces. Then follows a series of nucleon-nucleon collisions in which three nuclei are ejected, and two are kept inside thereby transferring some excitation energy to the "residual" nucleus. The crosses show the locations where dynamically possible collisions were forbidden by the Pauli principle

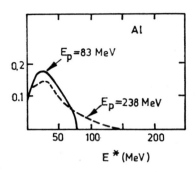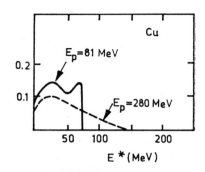

**Figures II-5, II-6**   Distribution of excitation energies in residual nuclei for two target nuclei at two proton bombarding energies. In ordinate, the fraction of residual nuclei with $E^*$ (from Metropolis et al.)

transparent to the incident particle, and the efficiency for energy retention decreases. The first step is best described as an intranuclear cascade (figure II-4): the incoming particle does not see the nucleus as a "black" body but as a collection of individual nucleons: indeed the de Broglie wavelength of the incident particle becomes smaller than the linear dimension of nucleus, and, at still higher energies, even smaller that the mean nucleon separation inside the nucleus. One moves from the domain of wave-optics to the domain of geometrical-optics.

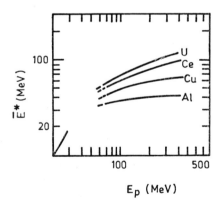

**Figure II-7**   Growth of the mean excitation energy $E^*$ in residual nuclei as a function of the incident proton energy for several target nuclei (from Metropolis et al.)

As a result of the intranuclear cascade, several particles are ejected from the nucleus, carrying with them a fraction of the incident kinetic energy. The excited structure consists in a distribution of "residual" nuclei (whose constitution depends on the number of ejected nucleons) with a distribution of excitation energies (examples are given in figure II-5 and II-6).

In figure II-7 the relation between the mean excitation energy and the incident kinetic energy is plotted.

The efficiency of energy retention, which is unity at low incident energies, is seen to decrease progressively as more and more energy happens to be transferred to the ejected particles. In the very high-energy range (GeV) the excitation energy becomes essentially independent of the incident energy.

In the second nuclear step (deexcitation) an increase in incident energy results in an increase of the number of possible channels and consequently

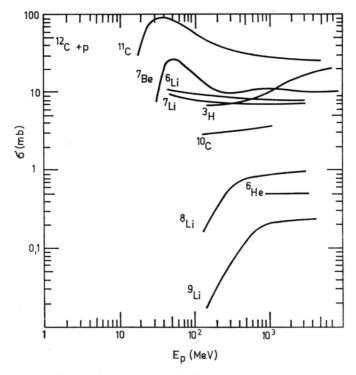

**Figure II-8**  Experimental cross-sections for the production of various isotopes from $p + {}^{12}C$

of possible final products. The branching-ratio toward any final product will clearly be related to the mass-difference between this product and the target nucleus, since the branching ratio is proportional to the volume of phase-space allowed to the corresponding final states, and since the volume of phase-space is itself a function of the available energies for these states, hence of the difference between the excitation energy and mass-difference ($Q$ value) between the target and final nucleus.

The volume of phase-space occupied by the sum of all reactions leading to a given final nucleus depends also on the number of bound states (and on their spin values) of this final nucleus. Furthermore, since the states containing this nucleus usually contain a certain number of light particles (nucleons, alphas, …) the probability of formation of this nucleus depends on the number of possible arrangements of these accompanying light particles.

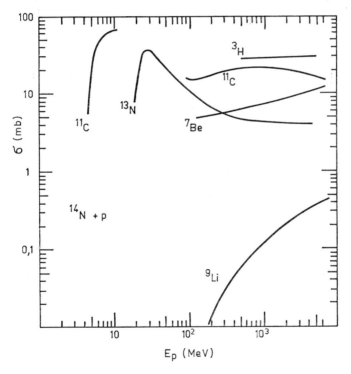

**Figure II-9** Experimental cross-sections for the formation of various isotopes from $p + {}^{14}\text{N}$

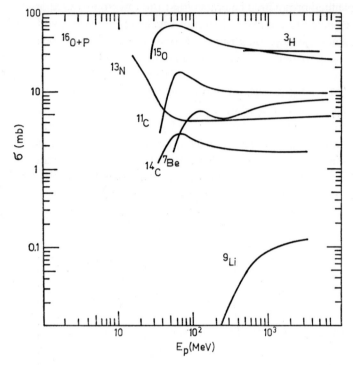

**Figure II-10**  Experimental cross-sections for the production of various isotopes from $p + {}^{16}O$

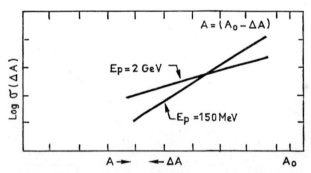

**Figure II-11**  If one plots the cross-sections for the formation of all products of given mass $A$ from a target $A_0$ ($\Delta A = A_0 - A$), one obtains a decreasing function of $\Delta A$ as shown in the diagram. The slope of the line decreases with increasing energy. Above a few hundred MeV the slope remains constant

The number of possible channels does not increase indefinitely; at very high energies it saturates. The saturation should take place when the (mean) excitation energy becomes a fair fraction of the total binding energy of the nucleus.

### II-1-3   Experimental results: the mass-yield*

In figure II-8, II-9, II-10 the experimental cross-sections for reactions leading from a given target to all measured products are plotted. The behaviour of these functions is studied qualitatively in figure II-11. Defining $\sigma(\Delta A)$ as the cross-section for formation of all nuclei of a given mass $A$ from a target $A_0$ ($\Delta A = A_0 - A$) we plot $\log \sigma(\Delta A)$ as a function of $\Delta A$ for fixed proton energies. The shape of the curves suggests a function of the form

$$\sigma \propto \exp[-p\,\Delta A] \qquad\qquad \text{(II-1)}$$

where $p$ (an empirical parameter) can be regarded as the inverse of the average $\Delta A$ reached at a given energy:

$$1/p = \overline{\Delta A} \qquad\qquad \text{(II-2)}$$

From figure II-11 $p$ is seen to be a decreasing function of the energy, leveling-off above a few GeV.

Numerically the empirical relationship

$$p = 0.1\,[E\,(\text{GeV})]^{-0.6} \qquad\qquad \text{(II-3)}$$

gives a reasonable fit in the largest part of the $A_0$ mass range.

When properly normalized to the total reaction cross-section, $\sigma(\Delta A)$ becomes:

$$\sigma(\Delta A)/\sigma\,(\text{reaction}) = p\,e^{-p\Delta A} \qquad\qquad \text{(II-4)}$$

This expression becomes rather inadequate (although not unduly so) in the low $\Delta A$ range. For large $\Delta A$ ($\Delta A \gtrsim A_0$, in particular if one considers the products T, $^3$He, $^4$He), this expression strongly underestimates the yield, as several of these light particles are sometimes emitted in the same nuclear event. As a rule, one should not apply it for $\Delta A > 2A_0/3$.

In figure II-12 the function $p\,e^{-p\Delta A}$ is plotted as a function of energy, using eq. II-3 for $p(E)$.

---

* This section and the following come mostly from the work of Rudstam.

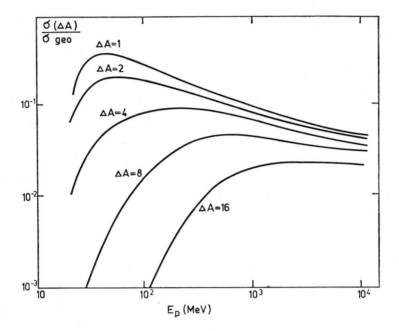

**Figure II-12**   Plot of function $p\,e^{-p\,\Delta A}$ (eq. II-4) with the empirical fit of $p$ (eq. II-3). Most of the well-known features of the experimental excitation functions are apparent; the smooth peak of the small $\Delta A$ reactions and their slow levelling-off at high energies; the slow, lazy rise of the large $\Delta A$ reactions (without any intermediate peak) to a high energy "plateau"

Some features of these curves are worth mentioning and discussing at this point.

First, we note that the cross-sections for small $\Delta A$ removal have a maximum at rather low-energy. As we move to larger $\Delta A$ the maximum shifts to higher energies and becomes broader and broader until it essentially disappears.

As we shall discuss later, this is related to the fact that the reactions leading to large $\Delta A$ removal are strongly endothermic (large $Q$ values). At low energy the deexcitation can only happen through a few channels, hence the relative probability for decay in any one of the channels is large. At higher energy new channels open up and take their share of the available phase space (the total cross-section is essentially constant). The result is a decrease of probability for the small $\Delta A$ channels. The new channels, however, come in an already crowded place and never manage to rise very high; in fact the very large $\Delta A$ values just smoothly rise to a flat plateau.

Second, we note that at very high energies the cross-sections all flatten up, reaching rather similar values. Indeed, the number of possible channels (toward given $\Delta A$ values) is, of course, limited: at these high energies no more new channel opens up and all existent channels share, more or less equally, the available phase space.

### II-1-4   Experimental results: the isobaric-yield

Next we look at the behaviour of the cross-sections for formation of products of the same mass (fixed $\Delta A$) but different isospin $T_Z = ((N - Z)/2)$. This is sometimes called the "isobaric-yield".

The measurements of $\sigma(T_Z)$ for fixed $\Delta A$ are appropriately represented by

$$\sigma(T_Z) \propto \exp - R\,(T_Z - T_{Zt})^2 \qquad\qquad \text{(II-5)}$$

where $R$ (an empirical parameter) is a decreasing function of the energy

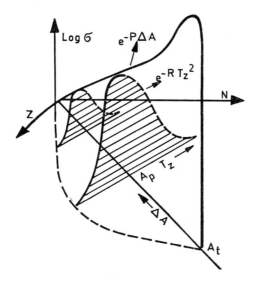

**Figure II-13**   Distribution of the experimental spallation yields (average over many experiments) in the $Z - N$ plane as a function of $\Delta A$ and $T_Z$ for a symmetric target of mass $A_t$ ($T_{Zt} = 0$)

($R \simeq 2$ around 100 MeV, and $\simeq 1$ around 1 GeV) and $T_{zt}$ is approximately the $T_z$ value of the target.

Hence we have altogether:

$$\frac{\sigma \, (A_0, Z_0, N_0 \rightarrow A, Z, N)}{\sigma \, (\text{reaction})} = \sqrt{\frac{R}{\pi}} \, p \exp \left[ -p \, \varDelta A - R(T_z - T_{zt})^2 \right] \quad \text{(II-6)}$$

The experimental results are decently matched by this formula at energies larger than a few tens of MeV.

The behaviour of this function is plotted in figure II-13.

### II-1-5  The experimental results as seen from the point of view of the statistical theory

The relationship between the nuclear statistical theory described previously and the experimental results is clearly seen through the behaviour of the nuclear binding energies as described by the mass formula.

As discussed in II-1-1, in a very coarse analysis, one finds that the mass difference between two nuclei lying at the bottom of the valley of nuclear stability (stable nuclei) is linearly proportional to $\varDelta A$ ($\simeq 8 \, \varDelta A$ MeV) or, in other words, the binding-energy is roughly 8 MeV per nucleon (figure II-2).

For nuclei of fixed $A$ and varying $T_z$ (isobars), the mass defect decreases as $T_z^2$ (due to asymmetry effects related to pairing energies, figure II-3).

The $Q$ value for the formation of a given product from a given target is clearly related to the mass difference and also consequently to the volume of phase-space allowed to this reaction.

The relation can be made a little more quantitative (although still in an approximate manner) If the final state involves $n$ particles, the most probable configuration is a sphere in $(3n - 4)$ dimensional momentum space (taking into account the conservation laws and the center of mass motion) Since the energies involved in the deexcitation are non-relativistic, $E^* \propto p^2$ and:

$$\sigma \, (A_0, N_0, Z_0 \rightarrow A, N, Z) \propto (E^* - Q)^{(3n-5)/2}$$

$$= \left[ E^* \left( 1 - \frac{Q}{E^*} \right) \right]^{(3n-5)/2} \alpha \, e^{-(aQ/E^*)} \quad \text{(II-7)}$$

where $a$ is an appropriate coefficient.

From the fact that $Q$ appears linearly in the exponent we expect (again through our approximate analysis of the mass-formula) the exponent to be linear* in $\Delta A$ and quadratic in $\Delta T_z$.

Furthermore, since $E^*$ appears in the denominator of the exponent in eq. II-7, we expect $p$ and $R$ in eq. II-5 to be (inversely) related to $E$. This is indeed the case: in the range 100 MeV to 1 GeV, both of these parameters are found to decrease, while above a few GeV they reach a constant value. Let us recall that in this higher energy region, the excitation energy does not increase any more with incident energy.

To summarize, we may say that the $\log \sigma$ surface in the figure II-13 represents a mapping of the mass formula (and, through it, of the properties of the nuclear forces) through the basic assumption of the statistical theories: the chain of reasoning goes from "probability of an event" to "volume of phase-space allowed to this event" to "energy available for the event" to "mass difference" to "mass formula" and to the fundamental nuclear properties involved.

## II-2  HIGH ENERGY NUCLEAR REACTIONS IN STARS

### II-2-1  The cross-sections for formation of the light elements; the formation ratios in stellar flares

#### II-2-1a  *The L-elements*

High-energy protons in stars can only be found in the surface layers (chromosphere or corona), where the gas densities are low enough to allow the operation of acceleration mechanisms (whatever they are). Flux of accelerated

---

* The alert reader will have noticed that the $\Delta A$ appearing in eq. II-1 should be larger than the $\Delta A$ predicted by eq. II-7 (as this last one is the difference between the mean residual nucleus and the final nucleus while the first one is the difference between the target nucleus and the final nucleus). This is the kind of (perfectly valid) objection that one tries to keep away from in a "simplified" version of a subject (as the answer usually takes one away from the desired simplicity).

Detailed calculations (Metropolis, Rudstam, etc...) have shown that the results of the cascade (the distribution of residual nuclei from a given target) and the results of the deexcitation (the distribution of final nuclei from a given residual nucleus) have quite similar $\Delta A$ and $\Delta T_z$ dependance. For instance, in a reaction emitting a total number of $\Delta A$ nucleons, about one third $(\Delta A)/3$ are emitted in the cascade and two thirds $(\frac{2}{3}\Delta A)$ in the deexcitation. As an example, in a reaction leading from $^{27}Al$ to $^{18}F$, the most probable residual nuclei have mass-24.

Hence the qualitative deexcitation behaviour predicted by eq. II-7 is not destroyed, but even strengthened by the cascade.

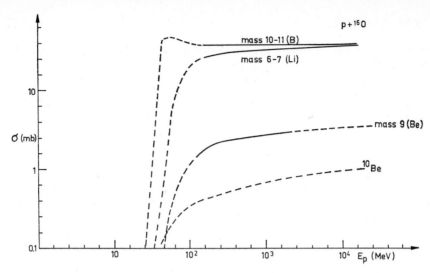

**Figure II-14**  Cross-sections (measured or estimated) for Li, Be, B formation by high-energy protons on $O^{16}$. The dashed parts are quite uncertain.

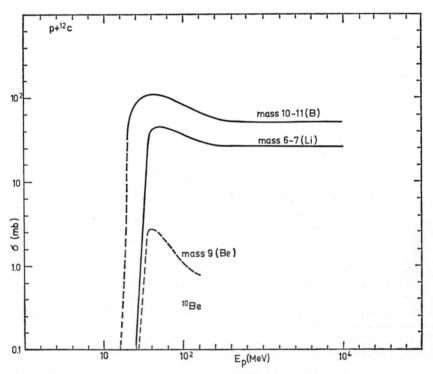

**Figure II-15**  Cross-sections (measured or estimated) for Li, Be, B formation by high-energy protons on $^{12}C$. The dashed parts are quite uncertain

protons are known to occur in our sun (solar flares), and presumably take place with varying intensities, in all stars.

Upon hitting the atoms of the stellar atmosphere, the fast protons will give rise to various nuclear products (spallogeneration) as described in section II-1. However, because of the scarcity of these fast particles (their number is severely limited by the amount of energy required, as discussed in section III), only very minute amounts of products can result. Hence the effect is best seen on very rare nuclei, such as $^6$Li, $^7$Li, $^9$Be, $^{10}$B, $^{11}$B and possibly also D and $^3$He. Because of their relatively high abundances, and because of the lower $Q$ values involved, the most important target nuclei in stars are the medium (M) nuclei $^{12}$C, $^{14}$N, $^{16}$O and $^{20}$Ne, with $^{12}$C and $^{16}$O being presumably dominant. Hence we shall discuss in the next paragraphs the probability of forming the L-elements (Li, Be, B) from the M-elements (C, N, O, Ne).

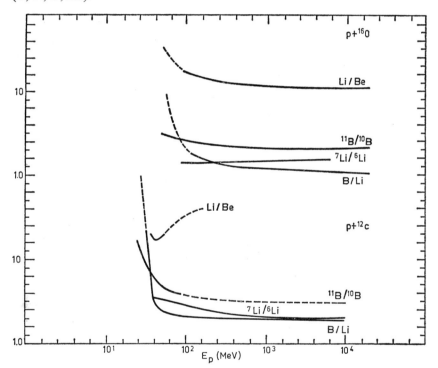

**Figure II-16** Ratios of the cross-sections for the formation of the L-elements by fast protons on $^{16}$O and $^{12}$C. The dashed parts are quite uncertain

The isotopes D and $^3$He would come mostly from $^4$He. They will be treated at the end of this section.

From our previous discussion, we should expect the cross-sections for the formation of the L-elements from the M-elements to be weakly decreasing functions of the mass difference ($\propto e^{-p\Delta A}$; $p$ as in eq. II-3). Hence we should expect, from the statistical theory $\sigma(Li) < \sigma(Be) < \sigma(B)$ for each target element M. The difference should be larger at low energies and should gradually decrease at higher energies.

However we must remember that our formulation of the statistical theory was based on a purely "average" behaviour of the nuclei ("smooth" mass-formula) and paid no attention to the special frailty of the mass 9 nucleus. Detailed Monte Carlo calculations of the nuclear reactions do indeed predict an especially small cross-section for this nucleus (Epherre and Gradsztajn).

Fortunately, for $^{12}$C and $^{16}$O most of the experimental data needed here is available (figures II-8, II-9, II-10), thanks mostly to the work of the Orsay group. In figure II-14, II-15 and II-16 the data for Li ($^6$Li, $^7$Li), Be ($^9$Be)

**Table II-1**  Best values (measured or extrapolated) for L-formation cross-sections (mb). Highly uncertain values are between brackets

| | | $p + {}^{12}C$ | | | $p + {}^{16}O$ | | |
|---|---|---|---|---|---|---|---|
| | | Li | Be | B | Li | Be | B |
| | | | | $\sigma$ (mb) | | | |
| $E_p =$ | 25 | (0.1) | | 64 | | | |
| (MeV) | 40 | 44 | 2.6 | 106 | (0.40) | | 20 |
| | 63 | 44 | (1.7) | 100 | (8) | (0.5) | 43 |
| | 100 | 37 | (1) | 80 | (20) | (1.2) | 40 |
| | 160 | 31 | (0.8) | 70 | 26 | 1.9 | 38 |
| | 250 | 26 | (0.8) | 58 | 29 | 2.2 | 38 |
| | 400 | 26 | (0.8) | 50 | 30 | 2.4 | 38 |
| | 630 | 26 | (0.8) | 50 | 32 | 2.6 | 37 |
| | 1000 | 25 | (0.8) | 50 | 32 | 2.7 | 37 |
| | 1600 | 25 | (0.8) | 50 | 33 | 2.8 | 37 |
| | 2500 | 25 | (0.8) | 50 | 34 | 3.0 | 37 |
| | 4000 | 25 | (0.8) | 50 | 34 | 3.2 | 37 |
| | 6300 | 25 | (0.8) | 50 | 36 | 3.3 | 38 |
| | 10,000 | 25 | (0.8) | 50 | 36 | 3.3 | 38 |
| Threshold (MeV) | | 23 | 26 | 16 | 25 | 32 | 23 |

and B ($^{10}$B, $^{11}$B) on $^{16}$O and $^{12}$C is summed and extrapolated (to the best of our knowledge) to the whole range of energy. The "best" data is tabulated in Table II-1.

Much less experimental data is available on the targets $^{14}$N and $^{20}$Ne. However, since these isotopes are usually less abundant than $^{16}$O and $^{12}$C we do not expect them to play a dominant role (unless the proton energy spectrum is highly concentrated in the low-energy range, in which case $^{14}$N would become important).

From Table II-1 and figure II-14 and II-15 it is clear that the ratios of the formation cross-sections are quite independent of energy above 80 MeV for $^{12}$C and 100 MeV for $^{16}$O. Taking a mixture of 0.33 $^{12}$C and 0.67 $^{16}$O (a common stellar mixture) we have:

$$\left.\begin{array}{l} \sigma(\mathrm{B})/\sigma(\mathrm{Li}) \simeq 1.5 \\[4pt] \sigma(\mathrm{Li})/\sigma(\mathrm{Be}) \simeq 20 \\[4pt] \sigma(^{7}\mathrm{Li})/\sigma(^{6}\mathrm{Li}) \simeq 1.7 \\[4pt] \sigma(^{11}\mathrm{B})/\sigma(^{10}\mathrm{B}) \simeq 3 \end{array}\right\} \text{above 100 MeV} \qquad \text{(II-8)}$$

From the table we find that as expected from the statistical theory:

$$\sigma(\mathrm{Li}) < \sigma(\mathrm{B})$$

$$\sigma(^{6}\mathrm{Li}) < \sigma(^{7}\mathrm{Li})$$

$$\sigma(^{10}\mathrm{B}) < \sigma(^{11}\mathrm{B})$$

but

$$\sigma(\mathrm{Be}) < \sigma(\mathrm{Li})$$

As mentioned before, this is an effect of the particular frailty of the mass 9 with respect to alpha-clusters. In terms of volume of phase-space, this frailty results in the absence of the $^{9}$B channel ($^{9}$B is particle-unstable), in the large $Q$ value for $^{9}$Be production (figure I-1), and in the fact that $^{9}$Be has only one bound state (the ground state).

In real life the fast stellar protons will not be monoenergetic, but will have a distribution of energies (energy spectrum).

By analogy with the solar flare protons we shall assume that these protons have a fairly steeply decreasing energy spectrum, expressible by a power law:

$$\phi(E)\,dE \propto E^{-\gamma}\,dE, \quad \gamma \simeq 2 \text{ to } 4 \qquad \text{(II-9)}$$

We notice from table II-1 that the ratio of the cross-sections do not vary much with energy. This will not remain true at very low energies however, because of the different $Q$ values (figures I-1, II-14, II-15). Nevertheless, recent data at 40 MeV (Davids, Epherre; this data is incorporated in the table II-1) strongly suggests that, unless the spectrum is exceedingly concentrated in the very low energy range, (corresponding to $\gamma \simeq 6$ or more) we should not expect the threshold energy region to matter too much.

We define, for any one of the L-elements, a probability of formation in a cosmic gas bombarded by a stellar flux of protons with spectrum $\phi(E_p)$

$$P_f(L) = \sum n(m) \int \phi(E_p)\, \sigma\, (m \rightarrow L, E)\, dE \qquad \text{(II-10)}$$

where $m$ is any one of the M elements, and $n(m)$ its relative abundance. Using eq. II-9 we see that $P_f(L)$ will be a function of $\gamma$, the slope of the spectrum. For comparison with observations it is more convenient to consider the ratios of the $P_f(L)$ for different L-elements.

For $(-\infty < \gamma < 1)$, (positive, zero or slightly negative slope) the ratios of the $P_f$ should be very much the same as the ratios of the $\sigma\ (E > 100\ \text{MeV})$ as given in eq. II-8. For more negative slope $(\gamma > 1)$ the ratios (in the order listed in eq. II-7) should all increase but not to a very large extent, unless $\gamma$ becomes appreciably larger than five or six. In Table II-2 limits and estimates of the important ratios are given.

Table II-2   Ratios of the important formation probability for L-elements. The "lower limit" is the ratio of the cross-sections above 100 MeV (eq. II-8). It would apply to energy-increasing or flat proton spectrum. The "upper limit" was computed for a spectrum with an energy power exponent of five $(\gamma = 5)$. Finally the column "estimate in stellar flares" applied to $\gamma = 3$ or 4. The nuclear data (measured or estimated) are those of Table II-1

|  | Lower limit | Upper limit | Estimates in stellar flares |
|---|---|---|---|
| $P_f(\text{Li})/P_f(\text{Be})$ | 20 | 40 | 30 |
| $P_f(^7\text{Li})/P_f(^6\text{Li})$ | 1.7 | 3.0 | 2.5 |
| $P_f(\text{B})/P_f(\text{Li})$ | 1.5 | 4.0 | 2.0 |
| $P_f(^{11}\text{B})/P_f(^{10}\text{B})$ | 3 | 6 | 4 |

We must also consider a few secondary causes of L-formation. Alpha-induced spallation reactions on CNO are likely to give quite similar L-ratios (both on theoretical and experimental grounds). Since the proton to alpha ratio in flares is usually large ($>7$) the effect is probably not important. The CNO induced spallations on $p$ and $\alpha$ (at rest) would naturally give again the same ratios. In absolute value the effect is probably similar to the effect of accelerated protons and alphas.

Reactions of $^4$He on $^4$He to form $^6$Li and $^7$Li could be of importance. See (Hayakawa) for a detailed account.

Finally the effect of secondary neutrons (mostly emitted from the $p + {}^4\text{He} \rightarrow n + \cdots$ reaction) is not important for any reasonably steep spectrum $\gamma \geqslant 2$. Otherwise the main effect is a decrease of the Li/Be ratio to $\sim 3$ (Audouze, Thesis).

## II-2-1b   The isotopes $^3$He, $d$, $n$

In table II-2b the cross-sections of proton-induced reactions on $^4$He are tabulated as function of energy (from Meyer, unpublished).

**Table II-2b**   Cross-sections of $p + {}^4$He for various events, and for the total yield of $^3$He, $d$, $n$ (after Meyer) (all in mb.)

| $E_p$ (MeV) | $\sigma$ (inel.) | $\sigma \begin{bmatrix} \text{He}^3 \\ + d \end{bmatrix}$ | $\sigma \begin{bmatrix} \text{He}^3 \\ + n \\ + p \end{bmatrix}$ | $\sigma \begin{bmatrix} ^3\text{H} \\ + 2p \end{bmatrix}$ | $\sigma \begin{bmatrix} d \\ + n \\ + 2p \end{bmatrix}$ | $\sigma \begin{bmatrix} 2d \\ + p \end{bmatrix}$ | $\sigma$ (He$^3$) | $\sigma(d)$ | $\sigma(n)$ |
|---|---|---|---|---|---|---|---|---|---|
| 23 | 0 | 0 | 0 | 0 | 0 | 0 | 0 | | 0 |
| 25 | 47 | 47 | 0 | 0 | | 0 | 47 | 47 | 0 |
| 28 | 60 | 46 | 5 | 9 | 0 | 0 | 60 | 46 | 5 |
| 30 | 70 | 45 | 13 | 10 | 0 | 0 | 68 | 45 | 13 |
| 35 | 93 | 44 | 31 | 13 | 2 | 2 | 88 | 50 | 33 |
| 40 | 107 | 43 | 42 | 16 | 2 | 3 | 101 | 54 | 47 |
| 45 | 116 | 40 | 47 | 17 | 7 | 4 | 104 | 55 | 54 |
| 50 | 119 | 38 | 49 | 18 | 10 | 5 | 105 | 58 | 59 |
| 60 | 120 | 33 | 49 | 19 | 12.5 | 6 | 101 | 57 | 62 |
| 70 | 118 | 27 | 49 | 19.5 | 14 | 7 | 96 | 55 | 63 |
| 80 | 113 | 21 | 48 | 20 | 16 | 7.5 | 89 | 52 | 64 |
| 100 | 104 | 13 | 46 | 20 | 17 | 8 | 79 | 46 | 63 |
| 120 | 95 | 7 | 44 | 20 | 17.5 | 9 | 71 | 43 | 62 |
| 150 | 86 | 3 | 40 | 20 | 17 | 9 | 63 | 38 | 57 |
| 200 | 80 | 1.5 | 36 | 20 | 16 | 8 | 58 | 34 | 52 |
| 250 | 78 | 1 | 34 | 20 | 15 | 7.5 | 55 | 31 | 49 |
| 300 | 78 | 0.7 | 33 | 20 | 15 | 7 | 54 | 30 | 48 |
| 400 | 82 | 0.5 | 32 | 24 | 15 | 7 | 56 | 30 | 47 |

The cross-sections listed represent, first, the probability of various events (for example $p + {}^4\text{He} \rightarrow {}^3\text{He} + n + p$) and in the last three columns the probability of forming one ${}^3\text{He}$, $d$ or $n$.

In figure II-16b the total destruction (inelastic) cross-section of ${}^4\text{He}$ and the cross-sections for formation of ${}^3\text{He}$, $d$, $n$ are plotted (from table II-2b).

In a cosmic gas the neutrons will either beta-decay or generate deuterium after proton capture (if the density is $\geqslant 10^{12}$ atoms/cm$^3$). At any rate, it is clear from figure II-16b that the ratio of $(d/{}^3\text{He})$ will be close to unity.

A meaningful comparison between the abundances of $d$, ${}^3\text{He}$ and Li, Be, B can be made with an assumption on the relative abundances of the target

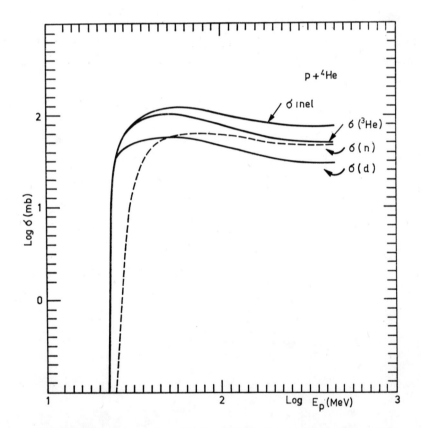

**Figure II-16b**  Cross-sections for interaction of $p$ and He$^4$. The total inelastic cross-section for formation of ${}^3\text{He}$, $d$, $n$ are plotted as a function of proton energies (J.P.Meyer)

atoms. If we assume that the irradiated stellar gas has a typical distribution of elements [$n(\text{H}) = 1; n(\text{He}) \simeq 10^{-1}, n(\text{CNO}) \simeq 2 \times 10^{-3}$] we could write, for instance, for the ($^3$He)/(Li) ratio

$$\frac{P_f(^3\text{He})}{P_f(\text{Li})} \simeq \frac{n(^4\text{He})\,\sigma\,(^4\text{He} \to {}^3\text{He})}{n(\text{CNO})\,\sigma\,(\text{CNO} \to \text{Li})} \simeq 200. \qquad \text{(II-11a)}$$

The other ratios can be obtained from Table II-2.

## II-2-2  The destruction cross-sections of L-elements

After their formation by nuclear reactions (commonly called spallation reactions) on the M-nuclei of the stellar surfaces, the L-elements can be partially or totally destroyed. The most likely reaction mechanism is the proton capture at low energies, leading to the destruction of the capturing nucleus. We have:

$$^6\text{Li} + p \to {}^4\text{He} + {}^3\text{He}$$

$$^7\text{Li} + p \to {}^4\text{He} + {}^4\text{He}$$

$$^9\text{Be} + p \to {}^6\text{Li} + {}^4\text{He}$$

$$\to 2\,{}^4\text{He} + d$$

$$^{10}\text{B} + p \to {}^7\text{Be} + {}^4\text{He}$$

$$^{11}\text{B} + p \to 3\,{}^4\text{He}$$

Each one of the channels is exothermic, hence opened to the lowest energy protons.

In figure II-17 the experimental cross-sections for the destruction of $^6$Li, $^7$Li, $^9$Be (and D) are given as a function of the proton incident energy. We note the following important points:

At very low proton energies (in fact at $E_p < 25$ keV) we have:

$$\sigma_d(^6\text{Li}) > \sigma_d(^7\text{Li}) > \sigma_d(^9\text{Be}) \quad E_p < 25 \text{ keV} \qquad \text{(II-11b)}$$

as expected from Coulomb effects.

At higher energies the $\sigma(^9\text{Be})$ increases faster than the others so that we have, for $E_p > 100$ keV:

$$\sigma_d(^9\text{Be}) > \sigma_d(^6\text{Li}) > \sigma_d(^7\text{Li}) \quad E_p > 100 \text{ keV} \qquad \text{(II-12)}$$

as expected from the frailty of $^9$Be.

Above 10 MeV the cross-sections probably become very similar.

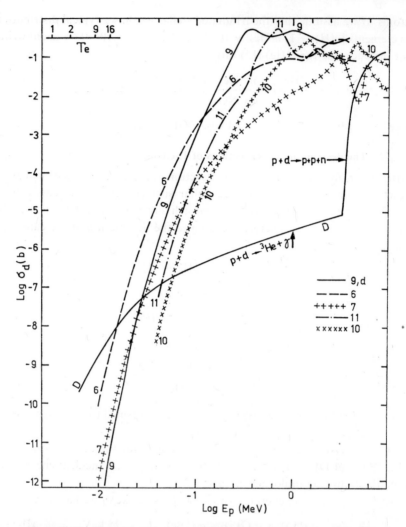

**Figure II-17** Experimental cross section for the destruction of D, $^6$Li, $^7$Li, $^9$Be, $^{10}$B, $^{11}$B as a function of incident proton energies. Note the very small scale of the ordinate. The insert in the upper left shows the temperature in million degrees for which the Gamow energies of Be are those of the abcissa. For example, at $10^7$ K the Gamow energy of Be (or Li or B, approximately) is $\simeq 10$ keV

For calculational purposes, the destruction cross-section $\sigma_d$ curves can be split into three regions: a low-energy Coulomb region, a "resonance" region (in the MeV range) and an absorption region at energies such that the destruction cross-section is approximately the geometrical absorption cross-section (many open channels).

We define $E_r$ the (experimentally determined) energy up to which the Coulomb effects appear to dominate the behaviour of the excitation function. In this region, one uses the Gamow formula, with a series development in power of energy of the nuclear $S$ factor

$$\sigma_d = (S/E)\,(1 + (S'/S)\,E + (S''/2S)\,E^2)\,e^{-2\pi\eta} \qquad E_p < E_r \qquad \text{(II-13)}$$

The energy is in the center of mass system and:

$$\eta = z_1 z_2 e^2 / \hbar v = (E_G/E)^{1/2}$$

$$E_G = (2\pi z_1^2 z_2^2\, Me^4) \simeq z_1^2 z_2^2 \quad \text{(MeV)} \qquad \text{(II-14)}$$

In the resonance region (from $E_r$ to $\simeq 10$ MeV) the cross-section is replaced by a mean value over the resonances $(\bar{\sigma}_r)$. This mean value is often found to be appreciably smaller than the absorption cross-section $\sigma_a$ as computed, for instance, from the optical model of the nucleus.

Above this region, at an energy $E_a$, again determined from the data, many channels open up and the destruction cross-section becomes essentially identical with the absorption cross-section. Hence:

$$\sigma_d \simeq \bar{\sigma}_r \quad E_r < E_p < E_a$$

$$\sigma_d = \sigma_a \quad E_p > E_a \qquad \text{(II-15)}$$

Experimental values are given in Table II-3.

**Table II-3** Experimental parameters for the destruction of D, $Li^6$, $Li^7$, $Be^9$. Also given, the Coulomb parameter $E_G$ defined in eq. II-14

| Element | $S$ MeV-barn | $S'/S$ MeV$^{-1}$ | $S''/2S$ MeV$^{-2}$ | $E_r$ MeV | $\sigma_r$ mb | $E_a$ MeV | $\sigma_a$ mb | $E_G$ MeV |
|---------|------|------|------|------|------|------|------|------|
| D | $3.6 \times 10^{-7}$ | 8.0 | 25 | 4 | 350 | – | 350 | 0.66 |
| $Li^6$ | 2.4 | $-0.8$ | 0.42 | 1 | 120 | 7 | 380 | 7.5 |
| $Li^7$ | 0.072 | $-0.5$ | 5.3 | 1 | 30 | 8 | 430 | 7.7 |
| $Be^9$ | 30 | 10 | 0 | 0.3 | 500 | – | 500 | 14 |
| $B^{10}$ | 9.12 | $-1.44$ | 2.5 | | | | | |
| $B^{11}$ | 100 | | | | | | | |

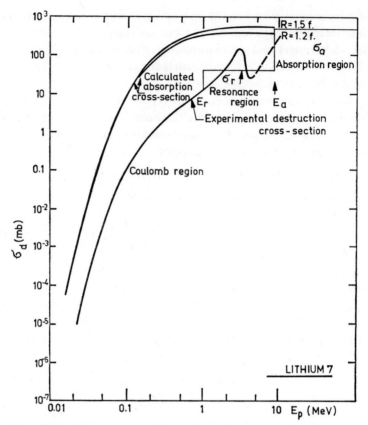

**Figure II-18**   The experimental destruction cross-section of $p + {}^7\text{Li}$ show-
ing the division in three regions; the low-energy Coulomb region, where the
approximation eq. II-13 is applicable (extending here to about 1 MeV);
the resonance-region where the actual values have been replaced by an
average value (eq. II-15); and the absorption-region where the destruction
cross-section (eq. II-15) becomes similar to the absorption cross-section
(computed here for two values of the nuclear radius)

The very low value of $E_r$ (0.3 MeV) for ${}^9\text{Be}$ and the fact that $\bar{\sigma}_r$ is already
as large as $\sigma_a$ shows again the particular frailty of ${}^9\text{Be}$.

In figure II-18 the specific example of ${}^7\text{Li}$ is described in details. The experimental destruction cross-section is plotted and the division in three
regions is shown.

The total absorption cross-section can be calculated from conventional
nuclear theory (Blatt and Weisskopf). The result is shown for two values of

the nuclear radius ($R = 1.2A^{2/3}$, $R = 1.5A^{2/3}$ in $f$). One does not expect to reproduce the absolute values of the cross-section in the low energy region (in part because the destruction cross-section neglects the decay of the excited nucleus by the entrance channel). At higher energies, many channels open, and the difference between the destruction cross-section and the absorption cross-section vanishes.

### II-2-3  Where do the destruction reactions take place?

To understand better the destruction processes we must have a closer look at the stellar conditions.

The high-energy protons in the stellar atmosphere can destroy some of the L-elements formed by their activity. While the L-elements were generated by protons of more than 20–30 MeV, they will be destroyed mostly by much lower energy protons ($\sim 1$ MeV) at least if the spectra are decreasing functions of $E_p$, as expected. The corresponding surface destruction probability can be written as:

$$P_d^s(L) = \int \phi(E)\, \sigma_c\,(L, E)\, dE \qquad \text{(II-16)}$$

where $\sigma_d\,(L, E)$ is the destruction cross-section of the element L (figure II-17). The most probable destruction energy $E_g^s$ (the energy for which the integrand has a maximum) is given, for a power spectrum*, by:

$$E_g^s = E_G/4\,(\gamma + 1)^2 \qquad \text{(II-17)}$$

and must, for all plausible spectrum, be situated above 100 keV and below a few MeV (for L-element). Hence we expect the relation II-12 to apply: in this region the $^9$Be abundance becomes the most sensitive test for the activity of fast (suprathermal) protons.

Yet another possibility exists for the destruction of the L-elements, which we should consider in some details.

From studies of stellar structures we learn that a large number of stars (T Tauri stars, Main-Sequence stars later than F2 or so, and others) have more or less extended surface convective layers. Through the convective motions occuring in these zones, the L-elements are carried down from the stellar atmosphere, and mixed all through the zone. The deeper the convective zone, the higher the temperature and the density in the bottom region and

---

* In more carefull treatment one would pay attention to the fact that the proton energy spectrum is usually referred to the lab system.

hence the higher the probability of thermonuclear destruction of the L-elements.

The probability of destruction in the convective zone is written as:

$$P_d^c(L) = \int n\, [E_p\, (\varrho,\, T)]\, v\sigma_d\, (L,\, E)\, dE \qquad \text{(II-18)}$$

where $n\, [E_p\, (\varrho, T)]$ is the Maxwell–Boltzmann distribution of hydrogen atoms at an appropriate mean temperature and density of the convective zone. Such integrals are familiar from studies of stellar interiors. Their integrands have a maximum at the Gamow energy defined by

$$E_g = (\tfrac{1}{2})\, (E_G^{1/2} kT)^{2/3} \qquad \text{(II-19a)}$$

$E_G$ is defined by eq. II-14.

Numerically $E_g \simeq 2.4 T_6^{2/3}$ for Li and $2.9 T_6^{2/3}$ for Be, where $T_6$ is the temperature in $10^6\,°K$ and $E_g$ the Gamow energy in keV. Clearly in all possible stellar cases of interest to us (surface convective zone) we have $E_g < 25$ keV. Hence the thermonuclear burning of the L-elements qualifies for the inequalities eq. II-11 i.e. $^6$Li will have the shortest lifetime, followed by $^7$Li and by $^9$Be. In figure II-19 the lifetime $t = 1/P_d^c(L)$ are plotted

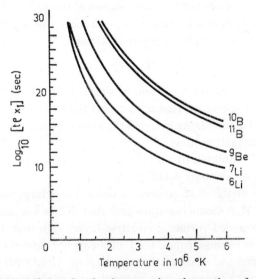

**Figure II-19**  Lifetime for the thermonuclear destruction of various isotopes, by proton reactions, as a function of the temperature. The ordinate is in reality the product of the lifetime, the density, and the fractional weight of hydrogen

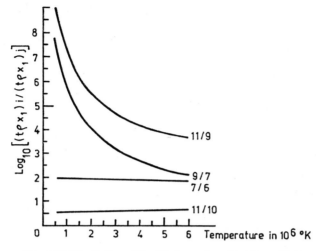

**Figure II-20**   Ratios of the lifetimes plotted in Figure II-19

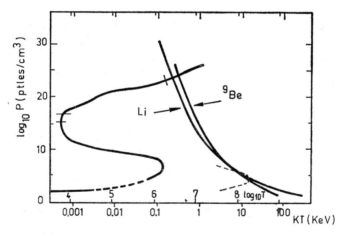

**Figure II-21**   In the density-temperature plane, the S-shaped curve represents the profile of the sun. The center of the sun is on the upper right. As we move away from the center, we meet first the bottom of the convective zone and then the optical layer (photosphere) where the temperature goes through a minimum. In the corona the temperature rises again and finally decreases in the interplanetary space. The dashed line in the bottom right is meant to represent the suprathermal component of particle believed to exist in the corona. The line "Li" shows the loci of the $\varrho - T$ points where $^7$Li burns in $\simeq 10^9$ years (and $^6$Li in $\simeq 10^7$ years). The line "$^9$Be" plays the same role for the destruction of $^9$Be. Note the crossing of the curves at $T \simeq 3 \times 10^7 \,^\circ$K (corresponding to a Gamow energy $E_g \simeq 25$ keV)

[actually the product of $t$ (the lifetime), $\varrho$ (the density in $g\,cm^{-3}$) and $X$ (the fractional hydrogen concentration)] as a function of temperature; and in figure II-20, the ratios of the lifetimes. The predominance of Coulomb effects at low energies is well exemplified by these figures.

In figure II-21 the existence of two different "regions" for the destruction of the L-elements is illustrated for the case of present sun.

The S shaped curve is the density-temperature profile of the sun. The center is at the upper right end of the curve. As we move outward we meet first the bottom of the surface convective zone, and later the optical layer where the temperature has a minimum. In the chromosphere, the temperature rises again. Furthermore a flux of suprathermal particles is expected to exist at least in certain locations. This second component is schematically pictured in the figure II-21.

On the same figure the locations of the $(\varrho, T)$ points at which the $^7Li$ burns in $10^9$ years (or $^6Li$ in $10^7$ years) and where the $^9Be$ burns in $10^9$ years are shown by two curves. The crossing of the curves around $T = 3 \times 10^{7\circ}K$ ($E_g = 25\,keV$) shows again the differential behaviour of Li and Be in the two burning sites.

### II-2-4   The time variation of the L-element abundances

The differential equation for the time variation of the L-elements can now be written as:

$$dn\,(L)/dt = P_f n\,(M) - P_d^s n\,(L) - P_d^c n\,(L) \qquad \text{(II-19b)}$$

Each term on the right-hand side of the equation is expected to have its own time-history. For instance the mean temperature in the convective zone (a parameter affecting $P_d^c$) rises gradually during the initial phases of the Hayashi contraction, goes through a maximum (figure IV-4) and decreases again as the bottom of the convective zone moves away from the stellar center, until it reaches a more or less constant value on the main-sequence (see however the possible effects of rotational breaking discussed in the next session). In later stages of stellar evolution, several break-up of convection to rather deep layers have been revealed through stellar model-calculations (see for instance figure IV-10).

The accelerated proton flux is also expected to have important time variations. The higher energy part (at least) is also believed to have been much more intense during the T Tauri phase and to be quieter during the main-sequence phase. On our sun, the activity in the tens of MeV range is concentrated in time and space, appearing in the form of sporadic flares. Recent studies strongly suggest the presence of a more or less continuous flux of ~MeV particles, as a kind of a background to the flares. Such a permanent activity of suprathermal particles is probably present in most stellar surfaces.

A thorough comparison of the observed abundances with theory would require a time integration of the various processes over the life of the star, taking into account the dilution of the L-elements in the convective zone. However some meaningful information can be obtained at much lower cost through a qualitative study of the behaviour of L-element ratios. Such a study will be the object of the fourth section of this course, in which we shall make use of the tools developed in this section.

# CHAPTER III

# Atomic Phenomena: The Energetics of L-element Formation

### III-1  ENERGY CONSERVATION AND ENERGY TRANSFORMATION

FROM THE LARGEST stellar Li/H and Be/H ratios (presumably observed in stars in which no destruction took place) we can obtain information on the amount of energy involved in the formation process. We must carefully distinguish two different meanings of the word energetics in the present context.

We may talk about the amount of energy lost in each spallation process (about 20 to 30 MeV, according to the $Q$ value). This energy is, of course taken from the proton kinetic energy. It represents altogether only a small expenditure for the star. For instance in a medium where $n(\text{Li})/n(\text{H}) \simeq 10^{-9}$ the actual loss is only 0.03 eV per nucleon of the medium: this first meaning belongs of course to the realm of the first law of thermodynamics ("you can't win").

We may also talk about the amount of energy that must have been transferred in fast particles (through some acceleration mechanism) if we take into account the fact that each accelerated-proton has only a minute chance of being credited for the spallogeneration of one L-atom. Indeed, by far the most numerous reactions that a fast proton will undergo are with the electrons of the medium; through its electromagnetic field this particle will accelerate the electrons in the vicinity of its orbit, thereby gradually loosing all its energy in the process. To give a specific example a proton of 60 MeV will be stopped after crossing $8 \times 10^{23}$ atoms cm$^{-2}$, or about one gram per square centimeter of ordinary stellar material. Since the number of appropriate targets (M nuclei) is about $n(\text{M})/n(\text{H}) \simeq 2 \times 10^{-3}$ and since the cross section for Li formation from an M nucleus is about $3 \times 10^{-26}$ cm$^2$, the probability $P_{\text{Li}}$ that this proton will generate an Li atom is

$$P_{Li} = n(\text{M})\, \sigma\, (\text{M} \rightarrow \text{Li}) \simeq 5 \times 10^{-5}$$

and the amount of energy required for the process is $\eta_{\text{Li}} \equiv E_p/P_{\text{Li}} \simeq 10^6$ MeV, or $10^4$ times more than in the first case. The important difference between the two cases is the fact that this energy is not lost for the star, but returns to the atmosphere in the form of heat (through a subsequent deceleration of the accelerated electrons themselves). Nevertheless this energy puts a severe drain on the stellar budget of energy transfer. Clearly we are here in the realm of the second law of thermodynamics ("you can't even break even").

## III-2  THE "STOPPING POWER" OF MATTER AND THE RANGE OF FAST PROTONS IN HYDROGEN

When a fast proton passes near an atom $i$ of charge $Z_i$ and energy spectrum $E_{ni}$ (including bound and free states) it can, through its electromagnetic field, excite the atom to one of these states $n$ thereby loosing an energy $E_{ni}$.

If $\sigma_{ni}$ is the cross-section for this event, then the total amount of energy loss by a proton crossing one cm of matter containing $N_i$ atoms (or $\sum_i N_i$, if one has a mixture) is

$$dE/dx = \sum_i N_i \sum_n E_{ni}\sigma_{ni} \tag{III-1}$$

It is convenient to express the path-length in unit of $dX = N_i\,dx$ (in atoms/cm$^2$); one defines:

$$\varepsilon \equiv -dE/N_i\,dx \tag{III-2}$$

the "stopping power" in MeV/atoms/cm$^2$. One also defines a path length in g cm$^{-2}$ (where $A_i$ is the mass number of atom $i$)

$$dX = \varrho\,dx, \quad \varepsilon = \frac{-dE}{\varrho\,dx} = \frac{-N\,dE}{A_i N_i\,dx}$$

where $N$ is Avogadro's number $(6.02 \times 10^{23})$.

The range of a particle of energy $E_0$ (in atoms/cm$^2$ or in g/cm$^2$) is defined as

$$X(E_0) = \int^{E_0} -dE/\varepsilon \tag{III-3}$$

### III-2-1  The case of a neutral gas

The eq. III-1 can be evaluated approximately by the following simplifications. An incoming proton interacts (independently) with each of the $Z_i$ electrons of the atom. The cross-section of interaction is proportional to $\pi\lambda^2$ where $\lambda = h/m_e v$ is the de Broglie wave-length of the proton-electron system (the motion of the electron around the atom is neglected). Upon being hit, the electron is imparted an amount of energy proportional to the ionization energy of hydrogen ($I = m_e e^4/2h^2 = 13.6$ eV; the natural energy of the problem). Hence:

$$\frac{-dE}{N_i\,dx} \propto \frac{\pi e^4 Z_i}{2m_e v^2} \quad \text{(for protons)} \tag{III-4}$$

If the incident particle has a charge $Z$ the cross-section becomes

$$\sigma \propto \pi z^2 \lambda^2$$

An accurate treatment (best done through the Born approximation) gives

$$-\frac{dE}{dx} = \frac{4\pi z^2 e^4 N_i Z_i}{m_e v^2}\left[\ln\frac{2m_e v^2}{I_i} + \ln(1-\beta^2)^{-1} - \beta^2 + \cdots\right] \tag{III-5}$$

where $\beta = v/c$ and $I_i$ is the mean energy of ionization of the atom $i$ (usually left as an adjustable parameter; the best values are 16 eV for hydrogen and 10 eV for heavier atoms). The term in bracket varies quite slowly with energy (less than a factor of two between 1 and $10^2$ MeV). A few useful approximations are listed here:

$$\frac{4\pi z^2 e^4 N_i Z_i}{m_e v^2} \simeq 0.301\beta^{-2}\,\frac{Z_i z^2}{A_i}, \quad \text{MeV g}^{-1}\,\text{cm}^{-2}$$

$$\simeq 143\,Z_i z^2/E\,(\text{MeV}), \quad \text{MeV g}^{-1}\,\text{cm}^{-2} \tag{III-6}$$

where $E$ is the (non-relativistic) energy per nucleon of the incident particle.

Also

$$\varepsilon = 340\,(Z_i z^2/E\,(\text{MeV}))\,[(\log E) + C] \quad \text{MeV g}^{-1}\,\text{cm}^{-2} \tag{III-7}$$

where $C = 2.14$ for hydrogen and 2.34 for heavier atoms.

Even more approximate but still very useful, one has:

$$\varepsilon = A/E\,(\text{MeV}) \quad \text{MeV atom}^{-1}\,\text{cm}^{-2} \tag{III-8}$$

with $A = 2\times 10^{-21}$ cm$^2$/atoms for protons in hydrogen.

From the fact that $\varepsilon$ for a particle of charge $z$ in a given material is simply $z^2$ times the $\varepsilon$ of a proton in the same material if the energy is expressed in MeV/nucleon, we get the following useful relation for the ranges

$$X_Z(E_0) = \int^{E_0} dE/z^2 \varepsilon_p\,(E/M) = \frac{M}{z^2}\int^{E_0/M}\frac{d\,(E/M)}{\varepsilon_p\,(E/M)} = \frac{M}{z^2}\,X_p\,(E_0/M) \tag{III-9}$$

### III-2-2  The case of an ionized gas (plasma)*

In the case where the atoms are entirely ionized the process is slightly different since the electrons do not start from the same initial state. Furthermore
we should expect the stopping power $\varepsilon_i$ for a proton of given energy to be a

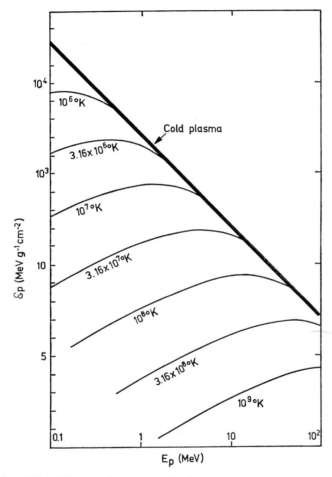

**Figure III-1**   The stopping power of hydrogen for protons as a function
of the temperature. For the cold plasma the approximation eq. III-12 has
been used

* This section is based on a treatment by L. Spitzer in *Physics of the Ionized Gas*.
Interscience Publishers Ltd, London, 1956.

decreasing function of the temperature of the plasma: in the limit of a very hot plasma the incident particle should not be decelerated (but only deflected).

One obtains:

$$\varepsilon_i = \frac{4\pi e^4}{m_e v_p^2} \ln \Lambda \left[ \psi(x) - x\psi'(x) \right] \tag{III-10}$$

where

$$\Lambda \simeq \frac{3}{2e^3} \left( \frac{k^3 T^3}{\pi n_e} \right)^{1/2} \; ; \quad x^2 \equiv \frac{m_e v_p^2}{2kT} \simeq \left( \frac{v_p}{\bar{v}_e} \right)^2 \tag{III-11}$$

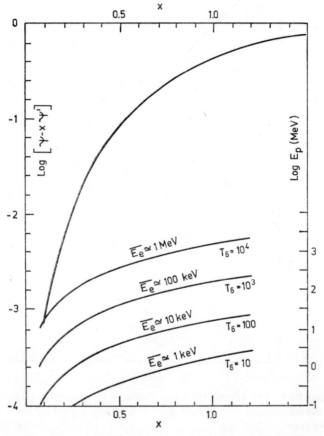

**Figure III-2**  Plot of the function $|\psi(x) - x\psi'(x)|$ in table III-1, as a function of $x$ (from Spitzer). In the lower part, the curves show the values of $x$ corresponding to a given choice of $E_p$ and $\overline{E_e}$

$\bar{v}_e$ being the mean electron velocity in the gas; $\psi(x)$ is the error function and $\psi'(x)$ its derivative with respect to $x$. The function $[\psi(x) - x\psi'(x)] \equiv S(x)$ is plotted in figure III-2.

For most stellar conditions of interest to us $\ln \Lambda \simeq 20$ (within 50%) and we find numerically:

$$\varepsilon_i \simeq 2900\, S(x)/E\,(\text{MeV}), \quad \text{MeV g}^{-1}\,\text{cm}^{-2} \tag{III-12}$$

with

$$E_p \simeq 0.16x^2 T_6 \tag{III-12}$$

In figure III-1 the value of $\varepsilon_i$ is plotted as a function of $E_p$ for a few temperature. (True thermodynamical equilibrium of the electrons needs, of course, not be realized; the "temperature" is a way of expressing the mean electron energies). From the figure III-1 it appears that $\varepsilon_i$ has a maximum around $x = 1$ and decreases rather slowly at $x < 1$. One interesting conclusion in our context is the following: in a medium with mean electron energies of $\simeq 100$ keV ($10^9\,°$K) the stopping power of stellar matter for the protons capable of spallation is reduced by a factor of ten or more over its value in a cold plasma. In the same fashion a medium of only a few million degrees will suffice to increase appreciably the range of the protons capable of L-element destruction (0.1 to a few MeV).

### Table III-1

| $x$ | $S(x)$ | $x$ | $S(x)$ | $x$ | $S(x)$ |
|-----|--------|-----|--------|-----|--------|
| 0.1 | 0.0007 | 0.6 | 0.132 | 1.2 | 0.59 |
| 0.2 | 0.0058 | 0.7 | 0.194 | 1.4 | 0.74 |
| 0.3 | 0.0193 | 0.8 | 0.266 | 1.6 | 0.83 |
| 0.4 | 0.044  | 0.9 | 0.34  | 1.8 | 0.91 |
| 0.5 | 0.081  | 1.0 | 0.42  | 2.0 | 0.95 |

### III-3   THE COMPUTATION OF THE ENERGY REQUIREMENT* L-ELEMENT FORMATION

We consider a proton of energy $E_p$ moving a distance $dX$ in a stellar† gas. Along this path the proton will loose an amount $dE$ of kinetic energy through

---

* This section is based on a paper by Ryter Reeves Graestajn and Andrage, entitled "The Energetics of light nuclei formation in stellar atmosphere" (to be published).

† As far as the stopping is concerned, we may consider only the hydrogen atoms, the other chemical elements in the stellar gas play an unimportant role.

electronic collisions. Along this same path $dX$ the proton has a probability $dP_L = \sum_m n(m)\, \sigma\, (m \to L)\, dX$ of generating an L-element. We define

$$\eta_L = dE/dP_L = \varepsilon_p n\, (H)/\Sigma n\, (m)\, \sigma\, (m \to L) \tag{III-14}$$

From the values given previously we have approximately, for a cold plasma

$$\eta_{\text{Li}} \simeq 8 \times 10^7/E\,(\text{MeV});$$

$$\text{for} \quad E \geqslant 30\ \text{MeV}, \quad \eta_{\text{Li}} \simeq 3 \times 10^6\ \text{MeV} \tag{III-15}$$

$$\eta_{\text{Be}} \simeq 1.2 \times 10^8/E\,(\text{MeV});$$

$$\text{for} \quad E \geqslant 35\ \text{MeV}, \quad \eta_{\text{Be}} \simeq 4 \times 10^7\ \text{MeV} \tag{III-16}$$

In a hot plasma the values of eq. III-15 and III-16 must be multiplied by the "thermal" factor $S(x)$.

# CHAPTER IV

# Stellar Observations and Interpretations

## IV-1   STELLAR OBSERVATIONS OF LITHIUM

THIS CHAPTER borrows a good deal from a recent summary by Wallerstein and Conti.

In figures IV-1 and IV-2 the abundances of Li/H are plotted in relation to the spectral class of the stars in which they are observed. The area assigned to a given group of stars are meant to represent the dispersion of the observed values in this group. Dashed curves mean that many "upper limits" are quoted: the real area may be much more extended.

As mentioned, the T Tauri stars have always high Li content ($\sim 10^{-9}$ or 3.0 in a conventional log scale in which $\log n(H) = 12$). It is worth remembering that T Tauri stars are believed to be very young stars, in their first period of contraction. Their age is probably less than $10^7$ years.

We notice next the cluster of the Pleiades; a rather young cluster with $T \sim 5 \times 10^7$ years. The values of Li are consistently high although perhaps a little less than for the T Tauri stars, especially in the case of the late G stars, where the difference reaches a factor of three or so.

In the somewhat older Hyades cluster ($T \sim 5 \times 10^8$ years) the decrease in Li abundance, as one goes to later type stars, is much stronger; a factor of ten or so appears between the early and late G stars. Three giant stars are observed in the Hyades, which have much less Li.

The "field stars" are expected to have a large range of ages. However in any reasonable model of galactic evolution, their mean age must be a fair fraction of the age of the galaxy.

Hence they should be in the average appreciably older than the Hyades. In the figure IV-2 the M.S.* field stars have their upper border along the Hyades area while their lower border extends all the way to the detection limit.

Next we consider the Giants and the Supergiants (field stars). In the F–G range they are more or less superimposed on the M.S.* field stars. For later type however we notice first a decrease; a minimum around K 5; and finally a strong increase in the M region. Most remarkable in this region are the

---

* M. S. stands for Main Sequence.

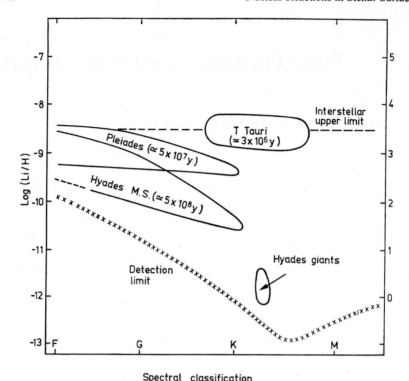

Spectral classification

**Figure IV-1**   The Li abundances observed in the T Tauri stars, the Pleiades and the Hyades are shown as a function of the spectral classes. The area assigned to a given group of stars are meant to represent the dispersion of the observed values in this group. Dashed lines are regions where upper limits are quoted. On the left scale appears the upper limit of interstellar abundances (Is). The line ($\times \times \times$) shows the detection limit. The scale on the right is a conventional log scale in which $\log n(H) = 12$ (adapted from Wallerstein and Conti)

Carbon stars where $n(\text{Li})/n(\text{H}) \simeq 10^{-7}$ are reported. However the abundance determination of Li in these late type is far from easy, and more accurate analysis are needed.

## IV-2   INTERPRETATION OF THE DATA (Lithium)

For pedagogical purposes it seems advisable to discuss the lithium data before presenting more observations.

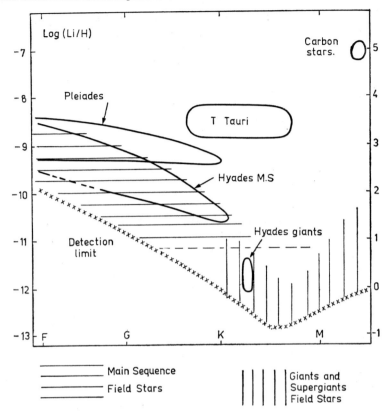

**Figure IV-2**   Same as figure IV-1, in which the area covered by Main-Sequence field stars and Giants and Supergiants field stars have been added

We start our interpretation by making the following fundamental hypothesis (due to Herbig): the stars are provided very early in their life with "initial" amount of L-elements as indicated in Table IV-1. No important addition takes place at least until the beginning of the "Giant" period.

We need not specify at this point whether these elements are of "galactogenic" origin or "autogenic" origin. This problem of origin will be treated in Section V.

The most widely accepted view today is the "autogenic" view: the T Tauri stars show strong indications of surface non-thermal activity (calcium emission features, strongly fluctuating light curve, intense mass loss etc...).* During this period the star would be the seat of very strong and intense

---

* This was written in March 1969.

**Table IV-1** Assumed "initial" stellar abundances of L-elements. The choice was guided both by the astronomical data and by the spallation ratios (Table II-2)

| | |
|---|---|
| $\log n(\mathrm{Li})/n(\mathrm{H}) = -8.9 \pm 0.5$ | (3.1) |
| $n(^{7}\mathrm{Li})/n(^{6}\mathrm{Li}) = \quad 2.5 \pm 1$ | |
| $\log n(\mathrm{Be})/n(\mathrm{H}) = -10.4 \pm 0.5$ | (1.6) |
| $\log n(\mathrm{B})/n(\mathrm{H}) = -8.6 \pm 0.5$ | (3.4) |
| $n^{11}(\mathrm{B})/n(^{10}\mathrm{B}) = \quad 3 \ \pm 2$ | |

"flares" in which the required spallation reaction would take place. As discussed shortly this phase must take place rather soon in the T Tauri period.

The deviation in the abundances of Li, Be and B is meant to represent the possible variation in the initial stellar abundances. As discussed in the first section the uncertainties in the formation ratios are smaller than a factor of three.

The statement that no important addition takes place during the main-sequence is strongly suggested by the similarity between the T Tauri and the Pleiades abundances, and by the fact that no M.S. stars (even those with shallow convective zone) are known to have Li abundance appreciably exceeding this "initial" contribution.

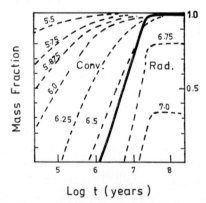

Log t (years)

**Figure IV-3** The internal structure of a star of 1 M$_\odot$ during its Hayashi contraction phase. The abcissa is the time since the beginning of the quasi-static gravitational contraction phase. The ordinate is the mass fraction from the center (bottom of the diagram). The dotted lines are isothermal lines, with $T$ in log. The solid line is the boundary between convective layers (at the left) and the radiative layers. (Adapted from von Sengbusch)

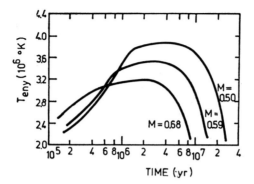

**Figure IV-4**   Temperature at the base of the convective zone as a function of time (on the Hayashi track) for three stellar masses (in unit of the solar mass). Significant lithium depletion in the envelope occurs when the temperature exceeds about $3 \times 10^6\,°K$ (from Bodenheimer)

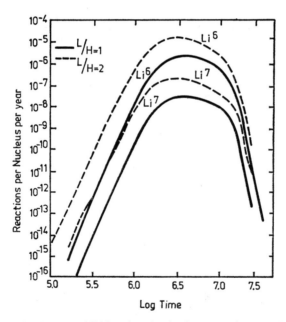

**Figure IV-5**   The rate of lithium burning in the convective zone for both isotopes versus their evolution time for two choices of the parameter L/H. (L/H is the ratio of mixing length to pressure scale height in the convective zone) (from Cameron)

Destruction of Li may occur in the Pre-Main Sequence period. During this period two different mechanisms influence the rate. At the beginning, the star is fully convective, but nowhere is the temperature high enough to cause any destruction. As the star moves down the Hayashi track (figure IV-3) the temperature increases everywhere. At the same time a radiative core is formed in the center and grows progressively in mass. As a result, the mean temperature in the convective region first goes up (because of the general warming-up) but eventually comes down (as the convective layers are pushed toward the surface). The maximum value reached is higher for smaller stars since these stars manage to have deeper convective zone all through this phase (this effect more than compensate the fact that, in view of their lower mass, they warm up at a slower rate). The evolution of the temperature and of the Li burning rate in the convective envelope* is described in figures IV-4 and IV-5. The corresponding evolutionary track is depicted in figure IV-6.

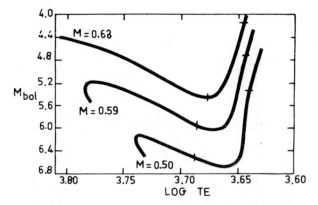

**Figure IV-6**  Pre-main-sequence evolutionary tracks for the three stars of figure IV-4. The weight composition which represents the Hyades mass-luminosity relation contains 38% of hydrogen, 60% of helium and 1.5% of heavier atoms; the $l/H$ ratio chosen here is 0.5. Lithium depletion in the envelope for each mass occurs primarily in the segment between the horizontal and vertical slash marks (from Bodenheimer)

The results of a detailed calculation are shown in figure IV-7 and compared with the observations of Li in the Hyades cluster. The calculation shows that Pre-Main-Sequence destruction of Li is important for stars later than G5

---

* I have not tried to see if the various figures given in these lessons are in very good agreement. They merely serve an illustrative purpose.

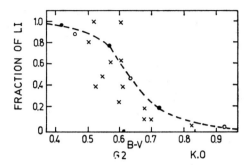

**Figure IV-7** The fraction of original lithium remaining in the convective envelope for contracting models that have reached the Main-Sequence plotted against arrival position on the Main-Sequence in terms of B–V (see Appendix). Open circles are for an hydrogen concentration of 64%; closed circles for 36%. The crosses indicate lithium observations in the Hyades (from Bodenheimer)

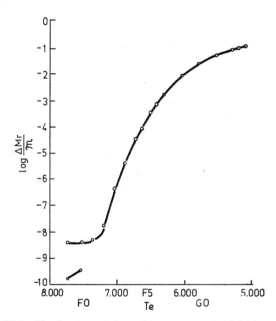

**Figure IV-8** The fraction of the stellar mass contained in the outer convection zone of a Main-Sequence star as a function of the surface temperature $T_e$ (from Baker)

or so, but very small for stars earlier than G2. Most of the observed stars in
the G0 to G5 region fall below the calculated curve. In the same fashion
Cameron (1965) has found that the Pre-Main-Sequence depletion of $^6$Li in
the sun is complete but about 60% of the $^7$Li manage to survive that period.
One could assign this discrepancy to further destruction on the Main-
Sequence (after all, the cluster has been on the M.S. for some $4 \times 10^8$ y).
However some difficulty still remains, as we shall discuss later.

Let us return to figure IV-1. Forgetting, for the moment, the quantitative
difficulty mentioned above, we shall assign the small depletion of Li
(noticeable if we compare the K Pleiades to the F Pleiades or to the T Tauri)
to Pre-Main-Sequence Li destruction, as described above.

Before we move any further it is interesting to make the following remark.
In figure IV-8 the fraction of the stellar mass contained in the outer convective zone of a Main-sequence star is shown as a function of the surface
temperature $T_e$. The fraction varies by $\sim 10^8$ between F0 and G0. From
figure IV-1 it is clear that no such strong correlation exist for the Li abundance
in the Pleiades. From this absence of correlation we may conclude a) that
either the Li is mixed throughout the star very early in the game, or b) on the
contrary the Li is confined to the photosphere and never really sinks all the
way down the convective zone. For reasons to be developed later we shall
prefer the first hypothesis. One corollary of this first hypothesis is that the

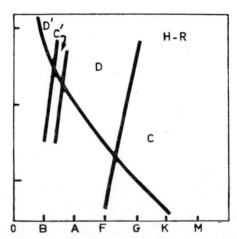

**Figure IV-9**  Regions of the H–R diagram where surface convective zones
are expected to be found. The narrow $c'$ band would be due to helium
ionization (from Schatzman)

generating reactions must occur very early in the T Tauri period, at a time when the star was still largely convective.

In figure IV-9 the H–R diagram is divided in regions in which stars have or do not have appreciable surface convective zone. According to Schatzman (private communication) a narrow band of surface convective zone (due to helium ionization) may exist in the region of the B stars.

Back to figure IV-1 a comparison between the Hyades and the Pleiades seems to indicate a further destruction of Li *during* the Main-Sequence evolution; the rather strong depletion of Li in advanced Hyades stars is in qualitative agreement with the fact that later-type stars have warmer convective zone and consequently destroy their Li faster than earlier-type stars. This long time-scale depleti of Li in the M.S. (first noted by Herbig) is of course reflected in the low mean value and wide dispersion of Li abundance in the M.S. field stars (figure IV-2).

The Li depletion on the M.S. is in qualitative agreement but quantitative disagreement with stellar models (Weynmann and Sears found no Li depletion for Hyades earlier than K 5). Several unsuccessful attempts have been made to bring the models in agreement with observations e.g. by taking into accounts special effects such as the overshooting of the convective cells in the radiative zone (Böhm–Vitense). The discrepancy appears to be real*.

Next we consider the Giants and Supergiants. The Hyades cluster contains some red Giant stars (figure IV-1) which, in view of the cluster age, must be rather massive stars, hence must have been rather early-type stars when they were on the M.S.

These stars show the same low Li abundance generally observed in field Red Giants earlier than K 5 (figure IV-2).

This low abundance is consistent with the calculations of Pre-Red-Giant stellar models (Iben, Kippenhahn and Weigert). Soon after leaving the M.S. a star will find its surface convective zone in the state of rapid expansion (figure IV-10). This temporary increase in the convective layer mixes the surface material with deeper material in which Li has been destroyed by thermonuclear reactions. One interesting case, in this respect, is the double star Capella with an F and G stars both belonging to the Giant branch. The less evolved component (F) shows at least one hundred times more Li than the more evolved component (Wallerstein). One reasonable explanation

---

* A promising attempt, based on a diffusion mechanism, is being made by Schatzman.

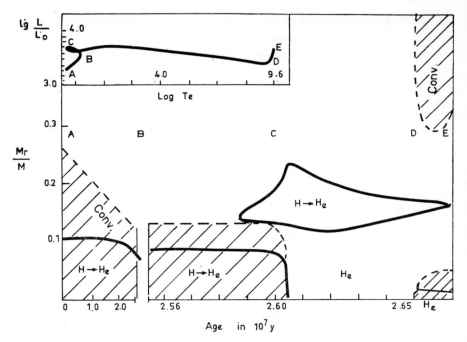

**Figure IV-10**   Internal structure of a $^7M_\odot$ through the hydrogen and the early helium burning phase. The areas surrounded by dashed lines represent convective layers; the areas surrounded by solid lines are nuclear burning regions. The sudden deepening of the surface convective layers is seen at the onset of the helium burning. The insert shows the corresponding path in the H–R diagram (from Kippenhahn)

would be to assume that the sudden deepening of the outer convection zone has already taken place for the G star but not for the F star. The corresponding path is shown in figure IV-11 (Iben).

| Point | Time | Point | Time | Point | Time |
|---|---|---|---|---|---|
| 1 | 0.024586 | 7 | 2.47004 | 14 | 2.55850 |
| 2 | 1.38921 | 8 | 2.47865 | 15 | 2.78295 |
| 3 | 2.23669 | 9 | 2.48429 | 16 | 2.94233 |
| 4 | 2.34089 | 10 | 2.48925 | 17 | 3.06968 |
| 4' | 2.34222 | 11 | 2.49817 | 18 | 3.19043 |
| 5 | 2.40119 | 12 | 2.50728 | 19 | 3.23566 |
| 6 | 2.44420 | 13 | 2.53163 | 20 | 3.26323 |

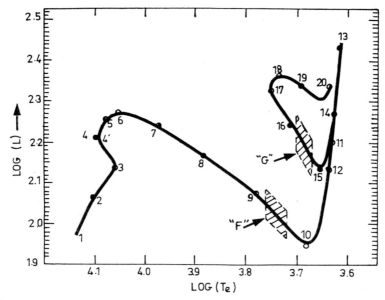

**Figure IV-11**   Path in the theoretical H–R diagram for a 3 M$_\odot$ population I star. The ordinate is in unit of solar luminosity ($3.86 \times 10^{33}$ erg/sec) and the surface temperature $T_e$ is in °K. The area F and G would represent the location of the two components of Capella. The evolutionary lifetime (in $10^8$ years) are given in the table

From figure IV-2 the Li abundance of giants and supergiants later than K 5 is found to increase tremendously. Several workers have expressed some doubts at the correctness of the abundance determination in these cold stars. If the determinations are correct we have here an indication of a renewal of spallo-generation (Feast) during the Red-Giant phase of stellar evolution. The similarity with the T Tauri phase comes to mind. In both phases we are in a region of deep surface convective zone (we are bordering the Hayashi forbidden region); in both phases we have important mass loss and impor-tant internal mechanical processes (structural readjustement). Spallogener-ation may find its place in this pattern.


IV-3   **CORRELATION WITH OTHER STELLAR PROPERTIES**

A number of important stellar features are found in the early F region of the main-sequence of the Hertzsprung–Russel diagram. They are first listed here:

A. The onset of the surface convective zone and its rapid deepening have already been mentioned. The phenomenon is illustrated in figure IV-8.

B. It has been known for quite a few years that M.S. stars earlier than F, systematically rotate faster than later type stars. In figure IV-12 the mean rotational velocities are seen to fall by a factor of ten in the range F0 to F9.

The maximum* possible speed of rotation (for the gravitational attraction just to balance the centrifugal force at the equator) based on the mass-radius relation ($M \propto R^2$) is, for a homogenous star

$$v \leqslant 360 \, (M/M_\odot)^{1/8} \quad \text{km sec}^{-1} \qquad\qquad \text{(IV-1)}$$

The early stars in figure IV-12 are only somewhat slower than the maximum value, but in the late F region the observed velocities are down by a factor of several tens.

C. Calcium emission lines are considered as witnessing chromosphere electromagnetic activity (they are seen in solar flares for instance). Along the

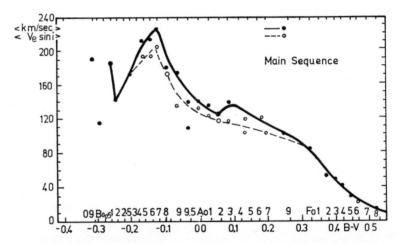

**Figure IV-12**   Rotational velocities as a function of spectral class for main-sequence stars. The notation $\langle v \sin_i \rangle$ means the average value of tangential component of the rotation velocity vector. (We have no way of knowing the direction of this vector). The open and closed circles represent two different ways of performing the statistics (including or excluding the $A_p$ and $A_m$ stars). (from Van den Heuvel)

---

* Much of what follows come from a number of papers by E. Schatzman.

**Table IV-2**   Properties of four groups of stars with different ages

| Object | T Tauri* | Pleiades G | Hyades G | Sun-field G |
|---|---|---|---|---|
| Age (y) | $5 \times 10^6$ | $5 \times 10^7$ | $5 \times 10^8$ | $5 \times 10^9$ |
| Ca II emis. | Strong | Strong | Weak | Weak |
| $\bar{v}_r$ (km/sec) | ? | 20 | 8 | 2 |
| Li/H | $1.5 \times 10^{-9}$ | $10^{-9}$ | $3 \times 10^{-10}$ | $3 \times 10^{-12}$ |

*In view of the similarity between the T Tauri stars we assume that their mean characteristics also apply to $\simeq 1\ M_\odot$ T-Tauri stars.

M.S. the Ca lines are reported (Kraft) to be absent before F 4, and to appear in some (but not all) stars of later type (figure IV-15).

D. In the range F 4–G the stars with Ca emission line turn faster than the stars without. In other words, the rotational velocity curve in figure IV-12 can, in fact, be split in two components, one above (with Ca II lines) and one below (without Ca II lines) the average curve (figure IV-13).

E. Conti has recently shown that amongst G stars those with $v_r > 3$ km/sec have more surface Li than those with $v_r < 3$ km/sec.

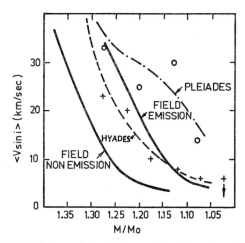

**Figure IV-13**   Variation of $\langle v \sin_i \rangle$ as a function of stellar mass (see appendix). Crosses and open circles correspond respectively to mean values for the Hyades and the Pleiades. "Field-emission" means M–S field stars with calcium emission lines; "field non-emission" means stars without calcium emission lines (from Kraft)

F. Finally, if we try to correlate all these parameters with age, we look at our four age land-marks: T-Tauri-Pleiades-Hyades-Field stars of about one solar mass (G stars in the M.S.). The results are given in table IV-2.

In figure IV-13 the mean rotational velocities of the Pleiades and the Hyades, field-emission and field-non-emission stars, are plotted as a function of spectral class (or equivalently of the stellar mass). The fastest rotators are always the Pleiades, followed by the field-emission stars and the Hyades and finally by the field-non-emission stars.

## IV-4  THE STELLAR DYNAMO

That the drop in rotational velocity would take place in the very range of spectral class (F0–F5) in which the surface convective zone becomes important is of course highly significant. The following pattern of interpretation has been suggested: a rotating convective zone is the seat of strong electromagnetic effect (dynamo); these effects are, in turn, responsible for particle acceleration, resulting in enhanced stellar winds and stellar flares. The star develops a complex and turbulent magnetosphere in which the fast particles must travel a long distance before they can really escape from the star. As a result the escaping particles carry with them a large angular momentum.

This way the rotation of the star is efficiently slowed down with the emission of a very small fraction of its mass ($<1\%$).

This model accounts for the decrease in rotational velocity in table IV-2 and figure IV-12: the deeper is the convective zone, the stronger is the effect and the faster is the deceleration.

This model also accounts for the behaviour of the Ca II line, since the chromospheric activity must somehow be related both to the rotational velocity and to the importance of the surface convective motions. In a sense the Ca II line intensity must be a measure of the rate of deceleration of the star; stars earlier than F are not decelerating because they have no convective zones; field non-emission stars are not decelerating because their rotation is already too slow to generate the strong magnetosphere and intense flares mentioned above.

In figures IV-14 and IV-15 some evolutionary tracks in the H–R diagram are given, for illustrative purposes. The relation with Ca II lines (H–K) is shown the figures IV-15.

The model, as such, fails to explain the Li behaviour (table IV-2). From the chromospheric activity in braking stars we might have expected an *increase* (not a decrease) in Li abundance with stellar age. However a closer look at the stability problem (raised by the fact that the angular momentum loss takes places at the surface of the star and may then generate a pattern of

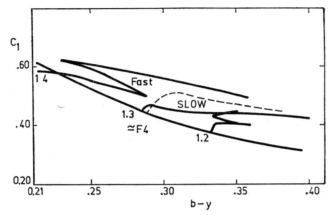

**Figure IV-14**   Evolutionary tracks for stars of 1.2, 1.3 and 1.4 $M_\odot$ in the H–R diagram. The composition is 67% H, 30% He and 3% heavier atoms. The dashed line is, according to Wilson, the boundary between the fast and slow rotators (from Demarque)

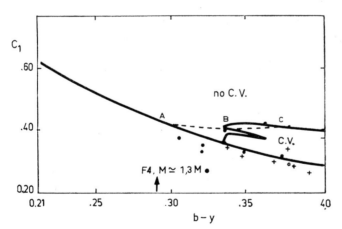

**Figure IV-15**   The dots, pulses and crosses represent stars with calcium emission lines of increasing intensities. Stars situated below the dashed lines have surface convective zone; stars situated above have none. The solid line is the evolution of a 1.2 $M_\odot$ star (from Demarque)

differential rotation) yields another clue for the understanding of the situation (Goldreich and Schubert, Spiegel). It has been argued that a stellar model in which the angular momentum per unit mass decreases somewhere with the distance from the axis of rotation is unstable to convection. This convection would bring surface material to inner layers (where the temperatures are higher than expected in the calculations described earlier) and would result in Li destruction.

Hence the stellar dynamo would influence the L-element abundances in two opposite ways:

a) the electromagnetic activity may increase the L-abundances by surface nuclear reactions.

b) the deepening of the convective zone may decrease the L-abundances by thermonuclear reactions. No calculation is yet available on the behaviour of the stellar dynamo. However from table IV-2 we know that the second effect should be the dominant one. We hope that this effect will help us in understanding the quantitative features of Li burning on the Main-Sequence stars (section IV-2).

## IV-5  STELLAR OBSERVATIONS OF THE $^7Li/^6Li$ RATIO

Determination of the ratio of $^7Li/^6Li$ have been reported for a small number of stars. Since the optical spectrum emitted by $^7Li$ and $^6Li$ slightly shifted one from the other, one can, in principle, obtain the abundance ratio from the line intensities. Recently several workers (Grevesse, private communication) have expressed serious doubts on the correctness of these determinations.

It is nevertheless conforting to find that the $^7Li/^6Li$ ratio all lies in the range from about 1.5 to larger values. Indeed from our nuclear physics analysis the formation ratio $P_f(^7Li)/P_f(^6Li)$ should be about 2 (Table II-2) and the destruction mechanisms (in the surface or in the convective zone; see figure II-17) can only increase this ratio.

The figure II-17 shows in fact that $^6Li$ is always (at all proton energies) destroyed much faster than $^7Li$. One would then expect the unaltered ratio of $^7Li/^6Li \simeq 2$ to occur in stars which have suffered no lithium depletion. This does not appear to be the case in stellar observations; the unaltered ratios are found mostly around G0 and *not* around F0. This problem will be discussed again together with the beryllium observations.

### IV-6  STELLAR OBSERVATION OF BERYLLIUM

The abundance of beryllium has been observed in a rather small sample of stars, extending mostly in the F0–G2 region. The results, for field M.S. stars are shown in the figure IV-16. There is a definite rise of $n(Be)/n(H)$ with spectral class: from F8 to G2 the value is about $3 \times 10^{-11}$ while around F0–F2 it is about $3 \times 10^{-12}$ or less.

In the same figure are shown the upper limit of the interstellar Be (barely higher than the highest observed stellar values) and the meteoritic value (normalized to H through the $n(Li)/n(Be)$ meteoritic value and the solar

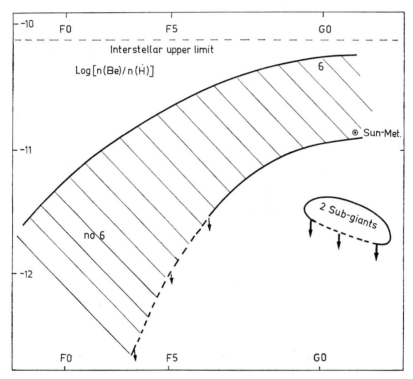

**Figure IV-16**  Distribution of beryllium abundances in Main-Sequence field stars; the shaded area shows the region where the points are situated, as a function of the spectral class. For some of the stars (those situated above the dashed parts of the lines) only upper limits are quoted. The notation "6" means that around this point some of the stars have the $(^{7}Li/^{6}Li)$ lithium spallation ratio ($\sim 2$); the notation "no 6" means that no such stars are found in this area (adapted from Wallerstein and Conti)

$n(H)/n(Li)$ value). As recently shown (by Grevesse and Hauge) the solar and meteoritic Be abundance are in good agreement.

In the figure IV-16 the notation "6" means that in the corresponding region a number of stars are observed to have $n(^7Li)/n(^6Li)$ close to the spallation ratio ("unaltered" ratio). The notation "no 6" means that no star are observed to have this property (i.e. $n(^7Li)/n(^6Li) \geqslant 5$ to 10). We note

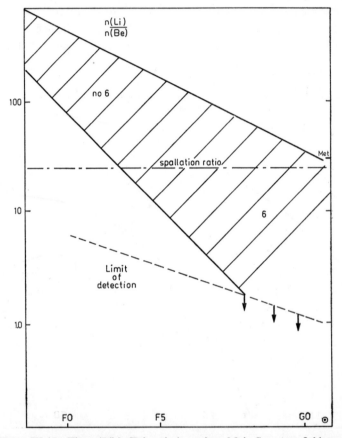

**Figure IV-17**   The $n$ (Li)/$n$ (Be) ratio in various Main-Sequence field stars (the same stars as in IV-16). The shaded area shows the region where the points are situated as a function of the spectral class. "Met" is the meteoritic ratio and---: the spallation ratio. The notation "6" means that around this point some of the stars have the lithium spallation ratio ($^7Li/^6Li \simeq 2$); the notation "no 6" means that no such stars are found in this area (adapted from Wallerstein and Conti)

that stars with low Be usually show "no 6" but stars with high Be quite often show the presence of $^6$Li.

In figure IV-17 the $n(\text{Li})/n(\text{Be})$ ratios are shown, in the same class range, for the stars appearing in figure IV-16. We note that stars earlier than F6 have an $n(\text{Li})/n(\text{Be})$ ratio clearly above the spallation ratio, (going up to about 5 or 10 times the spallation ratio in the F0 to F2 range). We note that stars later than F7 have, on the contrary, an appreciably smaller ratio. Finally we note again that the presence of $^6$Li (with a $^7$Li/$^6$Li close to the spallation ratio) is seen in the later type stars (F6 and later) but not in the earlier type.

## IV-7  AN INTERPRETATION OF THE COMBINED Li-Be DATA*

In section IV we have assumed that early in the T-Tauri phase the stars were given their share of L-elements, much in the spallation formation ratios discussed in the section II. The "initial" stellar amounts (normalized to H) are taken to be as in Table IV-1 where a deviation of a factor three each way is assumed to represent the variation from star to star.

From section II-2-3 we find that the best way of explaining the large Li/Be ratio (up to a few hundreds) in early F stars is the mechanism of L-element destruction by surface nuclear reactions (our $P_d^s$ of eq. II-16)†. That such large ratios should be seen only in stars without (or with very shallow) convective zone is perhaps more than a coincidence. The suggestion is that the Be atoms are destroyed in the F0–F3 stars on a longtime scale (after these stars have settled on the Main-Sequence) by a low energy (0.1 to 1 MeV) component of fast protons in a sort of residual electromagnetic activity. Such an activity may be analogous to the continuous acceleration of MeV protons in the surface of our sun.

In stars with an extended convective zone the L-atoms spend most of their time in deep layers in which they are effectively protected from the lethal surface protons. Numerically we may write

$$dn_L/dt = -n_L \sigma_d (L) \, \Phi \, (E > 0.1 \text{ MeV}) \, X \qquad \text{(IV-2)}$$

---

\* It is fair to warn the student reader that some of the ideas presented here are from recent cogitations of our own (Epherre and Reeves, to be published in *Astrophysical Letters*) and have not yet been submitted to the criticisms of other workers in the field.

† It is probably worth-while recalling at this point that recent measurements of spallation cross-sections in the low-energy range (Davids and Epherre, used in table II-2) have more or less ruled out an explanation based on a steep energy spectrum.

where $X$ is the ratio of the irradiated mass to the convective mass (assuming a thorough mixing). Hence

$$n_L(t)/n_L(0) = \exp\left[-\sigma_d(L)\, \Phi\,(E > 0.1\ \text{MeV})\, X\right] \qquad \text{(IV-3)}$$

From figure II-17 the average Be destruction cross-section is $\simeq 500$ mb at energies above 300 keV. When $X \simeq 1$ (early F stars) a flux of $\simeq 10^{25}$ ptle/cm² (integrated over the life of the star) would be required to appreciably destroy the Be.

For comparison, the sun (a notoriously weak dynamo) emits in the average $\simeq 5 \times 10^7$ protons of more than one MeV per cm² per sec, or $5 \times 10^{24}$ such protons per cm² since its arrival on the M.S. (assuming a constant rate). However, for the solar case, $X \sim 10^{-8}$ and the Be depletion must be entirely negligible.

Quite generally, as we move to later type stars, $X$ decreases rapidly (the irradiated mass remains essentially constant but the convective mass increases rapidly: see figure IV-8). This explains the rapid increase of the $n(\text{Be})/n(\text{H})$ ratio with spectral class (figure IV-16) till we reach the original ratio (table II-1) in the late F region.

This view explains, at the same time, three observational features:

a) The increase of the Li/Be ratio over the spallation ratio in the early F stars (by about one order of magnitude).

b) The decrease in the Be abundance $n(\text{Be})/n(\text{H})$ (from the "initial" values) in the same stars (by about the same value).

c) The absence of $^6$Li in these stars; indeed from figure II-17 we see that both $^6$Li and $^9$Be have a destruction cross-section which is at least one order of magnitude larger that of $^7$Li in the 0.1 to 1 MeV range.

In short, while the convective zone is "lethal" for Li, it is on the contrary a "refuge" for Be... Hence we are not surprised to find that the abundances of these two elements are inversely correlated with spectral class (figures IV-2 and IV-16).

Strong depletion of beryllium is observed in stars which have evolved past the Main-Sequence stage: Cepheids, supergiants and, already, in the two subgiants of figure IV-16. This is probably related (as for the lithium in the same stars (see section IV-2)) to surface dilution on the way to the "giant tip".

## IV-8  THE D/H AND $^3$He/$^4$He RATIOS IN STARS

A search for deuterium in stars of type B, A and F (Peimbert) has revealed only upper limits extending from $6 \times 10^{-4}$ down to $3 \times 10^{-5}$. In the sun, D/H $\sim 4 \times 10^{-5}$ (Kinman).

This is not surprising in view of the high rate of deuterium destruction in convective zone (figure II-17). In the pre-main-sequence phase, stars with 2.5 $M_\odot$ or less ($\simeq$ AO) are expected to destroy this element entirely, while in stars with $M > 3.5\,M_\odot$ ($\sim$ B6) it should be left intact (only one of the observed stars, $\gamma$ Peg, belongs to this second group: it has D/H $< 6 \times 10^{-4}$).

High ratios of $^3$He/$^4$He (larger than one) have been found in a few rather special stars which we have decided not to discuss here. Otherwise the upper limits are $\sim 10^{-2}$. Indirect indications (to be discussed next) give for the sun a value of $2 \times 10^{-4}$.

From our calculations of section II-2-1b, a proton flux, capable of generating the "initial" stellar abundances of table IV-1 by bombardment of a normal stellar gas, will yield:

$$D/H \sim 3 \times 10^{-7}$$

$$^3He/^4He \sim 3 \times 10^{-6}$$

From figure II-7 it is clear that the surface destruction of D by 0.1 to 1 MeV protons is negligible (this is also true for $^3$He and $^4$He because of the impossibility of radiative capture).

Thermonuclear destruction of D in convective zone will transform $D + p \rightarrow {}^3He + \gamma$. $^3$He will not be further processed.

Hence we have:

$$^3He/^4He \sim 6 \times 10^{-6} \quad \left.\begin{array}{l} \\ \\ \end{array}\right\} \quad M < 2.5\,M_0$$
$$D/H \sim 0 \qquad\qquad\qquad \text{(later than AO)}$$

$$^3He/^4He \sim 3 \times 10^{-6} \quad \left.\begin{array}{l} \\ \\ \end{array}\right\} \quad M > 3.5 M_0$$
$$D/H \sim 3 \times 10^{-7} \qquad\qquad \text{(earlier than B 5)*}$$

The low D/H upper limit are clearly not a problem. The isotope $^3$He is an important by-product of the hydrogen-helium conversion. Ratios of

---

* The possibility of a surface convective zone in B stars should be taken into account (figure IV-9).

$10^{-4}$ to $10^{-3}$ for the $^3$He/$^4$He are expected. Hence stellar $^3$He is most likely of thermonuclear origin.

## IV-9  THE SUN AND THE SOLAR SYSTEM

Of special interest, along these lines, is the study of our immediate environment, since a lot more information is available on the sun, the earth and the meteorites than on any other star.

It is widely believed that the solar system was formed very early in the evolution of the sun, most likely during its T Tauri phase.

Except for the small, and generally well understood, effect of the galactic cosmic rays, (and also of the long lived radioactive elements) the earth and the meteorites have not been the seat of nuclear reactions*.

Table IV-3   Abundance of L-elements in the solar system
In brackets the log scale described in the text

|  | Sun | | Meteorites* | |
|---|---|---|---|---|
| log [$n$(Li)/$n$(H)] | $-11.6$ | (0.4) | $-8.9$ | (3.1) |
| log [$n$(Be)/$n$(H)] | $-10.8$ | (1.2) | $-10.6$ | (1.4) |
| log [$n$(B)/$n$(H)] | $<-9.2$ | ($<2.8$) | from $-8.1$ | (from 3.9) |
|  |  |  | to $-9.8$ | (to 2.2) |
| $^7$Li/$^6$Li | $<20$ |  | 12.5 |  |
| $^{11}$B/$^{10}$B |  |  | 4 |  |
| (D/H) | $<4\times10^{-5}$ |  | $1.6\times10^{-4}$ |  |
| $^3$He/$^4$He | $<2\times10^{-2}$ | ($\simeq2\times10^{-4}$) | $\simeq2\times10^{-4}$† |  |

* The meteoritic ratio are "normalized" to H through ratio of H/Si = $10^{4.45}$ (Grevesse).
† Excluding the effect of galactic cosmic rays since the meteorites have solidified.

In this sense we should find here a memory of the state of the solar nebula some five billion years ago (although we have to be aware of possible geochemical fractionation of meteoritic elements, and even isotopes).

Consequently, a comparison between the solar photospheric abundances and meteoritic abundances should essentially give us two points in a time-sequence-evolution of the alteration processes.

In the table IV-3 a comparison is presented between solar and meteoritic

---

* At least until very recently...

data. It is of interest to note that the Be abundances agree quite well (although we should keep in mind that the uncertainties are large); indeed according to the views presented previously we expect no important formation or destruction of this isotope in G2 stars after the initial contribution.

The agreement between solar and meteoritic Be/Si has another important implication: the spallogenerating reactions took place before the separation of solar and planetary matter as, otherwise, the geometrical factors would be likely to destroy this agreement. This ties in with the idea that spallogeneration took place very early in the T Tauri phase.

The Li/Be ratio in meteorites ($\simeq 50$) is in reasonable agreement with the cross-section ratios at fixed energies in the range 40 MeV on up ($\simeq 25$). As discussed before a small increase is expected in the energy-averaged cross-sections, because of the contribution of the lower energy range. (We have accordingly chosen the value of 30 for the "Spallation ratios".) One implication is that very little Li destruction has taken place between the moment of formation of the light isotopes and the formation of the meteorites. This is coherent with the ratio of $^7Li/^6Li$ in meteorites. This ratio ($\simeq 12.5$) shows some depletion of $^6Li$. It is not clear at the present time that this depletion was caused by a nuclear process (the question will be discussed later).

If it were then, in view of the fact that the $^6Li$ destruction cross sections is, at all energy, much larger than the $^7Li$ destruction cross section (figure II-17) one can easily see that the total lithium abundance cannot have decreased by more than 40% in the process in which the $^7Li/^6Li$ ratio increased from 2.5 to 12.5.

In the surface of the sun the total lithium abundance has decreased by a factor of $\simeq 500$. This means that the average destruction life time of $^7Li$ is about $7.5 \times 10^8$ years*, requiring a temperature of $\simeq 2.8 \times 10^6$ °K in the convective zone (figure II-19). As mentioned before, this value is higher than the value predicted in stellar models of G2 stars ($< 2 \times 10^6$ °K), but may possibly be explained by the rotational braking mechanism (IV-4).

The meteoritic B/Li ratio has recently been found to vary from $\sim 7$ to 0.2, while the cross-section ratio is $\simeq 2$. Furthermore the meteoritic B/Li ratio is correlated with the petrographical properties of the meteorite in which it is measured, possibly suggesting an enrichment of B for the high ratios and a depletion of B for the low ratios. At any rate we have here a good example of alteration of L-element abundances by geochemical effects after the

---

* Recent result on two other clusters have confirmed this value (Danziger).

nuclear processes. This example of course teaches (1) that meteoritic element abundance ratios should be treated carefully (2) that, nevertheless, there are often ways of untangling the various alteration processes.

No boron has yet been detected in the solar surface (or in any other stellar surface). From table IV-3 the observational upper limit is $n(B)/n(Be) < 40$. This number can be directly compared to the ratio of the formation probabilities ($\simeq 60$, in table II-2) since we expect no destruction of these elements in the solar surface or in the solar convective zone. In view of the uncertainties the difficulty is not serious.

To pursue our discussion we note that the $^7Li/^6Li$ ratio in meteorites (12.5) is certainly larger than the formation ratio ($\simeq 2$). Therefore some depletion of $^6Li$ must have taken place. Furthermore the fact that this ratio is very much the same in the earth and in the various meteorites implies that the depletion occurred before the accretion of these solid bodies hence probably while this material was still in a gaseous form (possibly including some dust) as, then, homogeneity is much easier to realize (through mass motion). Several hypotheses have been presented to account for the meteoritic depletion of $^6Li$. They are described briefly here:

A. By analogy with the stellar cases discussed earlier (section IV-2) we may assume that the lithium-6 depletion occurred in the convective zone of the early sun. But we must realize that we are then buying two other important assumptions:

   1) the material of the inner solar system was previously in the sun itself, and was, later, ejected to its present place,

   2) the formation of the solar system took place quite late in the T Tauri period; during most of this period the sun was indeed not hot enough to burn $^6Li$ (see figure IV-3, IV-4 and IV-5).

Furthermore this hypotheses meets with one important difficulty: the presence of D (D/H $\simeq 1.6 \times 10^{-4}$) in the solar system. It will be quite clear from figure II-17 that at thermonuclear reaction energies, any burning of $^6Li$ implies a complete burning of D. Special assumptions would have to be added to explain the meteoritic D. Although none of these assumptions can be rejected at once, nevertheless, as always in scientific enquiries, they weigh against the model.

B) Alternately, we may assume that $^6Li$ was destroyed by nuclear reactions in the solar nebulosity itself by suprathermal protons. From figure II-17 we see that, in the range $10 < E_p < 100$ keV, $^6$Li is the most vulnerable light element (including even deuterium).

C. Finally, we must consider non-nuclear effects. At a certain stage of its evolution the inner solar system has lost most of its hydrogen and helium (together with most of its CNO Ne). We do not know how this loss took place. A simple evaporation process has been suggested long ago, but is now known to run into severe difficulties, and cannot be solely responsible for the loss of the volatile elements of the inner solar system. Nevertheless this process could have played a partial role in the loss phase.

Such a process favors of course the loss of light nuclei (as, at a given temperature, they have the highest velocity) and would, as a result, tend to increase the $^7$Li/$^6$Li ratio. However, to account qualitatively for the observations, rather specific conditions of temperature and densities are required, and before a realistic model is worked out, no conclusion can be drawn.

D. A somewhat "ad hoc" mixture of "big bang" contribution with spallogeneration has been proposed (Mitler). The difficulties attached to the various classes of theories will be discussed in section V.

As a result it is fair to say that we cannot offer a convincing description of the mechanism by which the lithium isotopic ratio in the solar system took its present value.

We come next to the $^{11}$B/$^{10}$B ratio. Again the earth and meteoritic ratio are very much the same ($4 \pm 0.4$) and, this time quite compatible with the spallation ratio on a cosmic gas ($4 \pm 2$). From figure II-17, we see that no alteration should be expected, either in the hypothesis of a passage through the solar convective zone or in the nebulosity itself (unless mean energies of $>1$ MeV prevailed—in which case $^{11}$B would have suffered some destruction—a rather unlikely situation). In fact the $^{11}$B/$^{10}$B ratio constitutes one of the best proofs of the spallogenesis of the light elements.

Finally we consider the D/H and $^3$He/$^4$He ratios. The D/H ratio has been measured in meteoritic water as well as in the earth. The ratio is apparently very much the same ($\simeq 1.6 \times 10^{-4}$) and somewhat larger than the solar upper limit ($< 5 \times 10^{-5}$). The implication is that the solar system matter has not been involved in the Pre-Main-Sequence depletion of solar deuterium, hence the separation of the sun and the nebula must have taken place before deuterium burning (at about $10^6$ °K).

From the discussion of IV-7b the meteoritic D/H ratio is far too large to be explained by spallogenesis alone. One likely explanation is in term of isotopic concentration of D/H during the escape of the volatile gases from the inner solar system. (The same as our explanation C) for the $^7Li/^6Li$ ratio).

The $^3He/^4He$ ratio in meteorites ranges from $2 \times 10^{-4}$ to $\simeq 0.1$. The spallation of meteoritic atoms by galactic cosmic rays after the solidification of the meteorites is responsible for the high ratios. The lowest value $(2 \times 10^{-4})$ is generally taken as reflecting the abundances in the solar nebula.

This same ratio has been found also in the "gas-rich dark portion" of a number of meteorites (Pepin and Signer). The discovery of large track densities in microscopic crystals (Pellas et al., Lal) shows that the gas enrichment must be due to solar emitted energetic particles (wind and flares) in the early days of the solar system (Eberhardt et al.). Consequently, over-looking possible alteration effects due to the acceleration process (the $Z/A$ ratio are quite different) we may assign this value to the early solar photo-sphere*. As discussed before this ratio is in the range of the thermonuclear formation ratio at the end of the hydrogen-burning-stage. In fact the ratio may be even too small; values of $10^{-3}$ would be expected (W.A. Fowler, private communication). Preferential evaporation may be needed again.

In short the meteoritic matter seems to present a fair example of the spallation products arising from the bombardment of a cosmic gas by energetic protons: the Li/Be and the $^{11}B/^{10}B$ ratio are in reasonable agree-ment, while the B/Be agreement is more than suggestive (despite the spread of values). Although the $^7Li/^6Li$ ratio does raise a problem, the difference is still within an order of magnitude and, in this sense, points again toward a spallative origin.

The problems left are with the D/H, $^3He/^4He$ and $^7Li/^6Li$. The "initial" stellar values are $\simeq 6 \times 10^{-7}$, $\simeq 10^{-3}$ (mostly due to the hydrogen burning stage in previous stars) and 2. It is interesting to note that the observed values in table IV-3 all show an increase of the heavier component of each ratio and that this increase is larger for the lighter atoms ($\simeq 300$ for D/H; $\simeq 10$ for $^4He/^3He$; $\simeq 5$ for $^7Li/^6Li$). We note that these alterations are indeed in the direction expected in the case of an evaporating gas.

Two other observations are of interest in this respect.

1) A recent determination of $^{12}C/^{13}C$ in the sun (Lambert) gives $\simeq 130$ while the terrestrial value is 90.

---

* I am endebted to P.Pellas for this remark.

**Table IV-4**   A comparison between meteoritic ratios and the early sun ratios determined (and sometimes guessed...) by methods discussed in the text. All ratios are increased in a direction favoring the heavier atoms. The increase is largest for the lighter atoms

| Isotope ratios | "Early sun" | Meteoritic | Increase |
|---|---|---|---|
| D/H | $6 \times 10^{-7}$ | $1.6 \times 10^{-4}$ | $\simeq 200$ |
| $^4$He/$^3$He | $\sim 10^3$ | $5 \times 10^3$ | 5 |
| $^7$Li/$^6$Li | 2.5 | 12.5 | 5 |
| $^{11}$B/$^{10}$B | $4.0 \pm 2$ | 4 | $> 50\%$ |
| $^{13}$C/$^{12}$C | 1/130 | 1/90 | $\simeq 35\%$ |
| $^{22}$Ne/$^{20}$Ne | $\simeq 1/14$ | 1/10 | $\simeq 40\%$ |

2) The gas-rich meteorites show a ratio of $^{20}$Ne/$^{22}$Ne about 40% higher than carbonaceous chondrites. Attributing, as before, the gas enrichment of meteorites to an early solar irradiation, we are led again to an enrichment of the solar system in the heavier atoms (as compared to the sun). The data are summarized in table IV–4.

Although the "early sun" determinations are indirect in nature and rest sometimes on highly speculative hypothesis nevertheless the apparent regularity may point out toward the operation of an evaporation mechanism in the early days of the solar system*.

In 1962 Fowler, Greenstein and Hoyle have proposed a model in which neutron capture on cold planetesimals played a crucial role in altering the formation ratios of L-elements in meteorites. Since then, the accurate determinations of the cross-sections (table II-1) have indeed shown that no alteration is needed for the boron ratio (and that, in fact, the agreement discussed in the previous paragraph could be used as an argument against an important neutron flux). This model has also met with other difficulties which make it rather unplausible at the present time. However this model has been very fruitful in that it has inspired a large number of workers in various fields and, in that way, has, so far, played a very useful role in the development of the subject.

## IV-10  THE STELLAR ENERGY PROBLEM AND ITS IMPLICATIONS

In section III-3 the amount of energy $\eta_L$ needed to form the L-elements was discussed for the case of cold and hot plasmas. It was found that, for mean

---

* The student reader should be warned of the fact that this last speculation is very new and yet uncritized...

electron energies ($\bar{E}_e$) less than $\simeq 20$ keV one required $\simeq 2 \times 10^6$ MeV (3 ergs) per lithium atom or $4 \times 10^7$ MeV (60 ergs) per beryllium atom. At higher mean electron energies, this requirement was somewhat reduced but, even for $\bar{E}_e \simeq 100$ keV, the reduction hardly reached a factor of ten.

The energy problem for the star as a whole can be considered in two different way.

We may argue that, since the young stars have Li photospheric abundances of $[n(\text{Li})/n(\text{H})] \simeq 10^{-9}$, and since this ratio is likely to prevail in a large part of the mass (because of the convective motions) the energy requirement per nucleon is:

$$\eta^* = \eta_L \times \left( \frac{n(L)}{n(H)} \right) \qquad \text{(IV-4)}$$

This quantity is about 3 keV in a cold plasma and (as discussed before) can hardly be expected to decrease by more than a factor of ten in realistic situations.

This quantity is, at any rate, of the same order as the total gravitational energy of the star. Since the T Tauri stars have (supposedly) not yet burned any nuclear fuel, they must have lived and shone entirely on their gravitational energy release. Hence, if the atoms of lithium were formed in situ (autogenic view), an important part of the luminosity of a contracting star must be emitted in the form of high energy particles.

One could, of course, challenge the assumption of lithium homogeneity in T Tauri stars, and pretend that, in view of the very strong mass loss, the L-elements are not effectively retained by the star during this period.

Another approach (due to Poveda) circumvents this objection and brings us to the same conclusion. In T Tauri stars, the rate of mass loss is known to reach values of $10^{-6}$ $M_\odot$ per year. In an atmosphere with $[n(\text{Li})/n(\text{H})] = 10^{-9}$ this corresponds to the loss of about $3 \times 10^{34}$ atoms of lithium per second. If we assume that an equilibrium has been reached between the formation rate (by spallation reactions) and the loss rate, we are led to a "particle luminosity" "$L_{\text{part}}$" of $10^{35}$ erg/sec; quite comparable to the optical luminosity of the T Tauri stars, hence to the instantaneous release of gravitational energy*.

---

* It is fair to say that the very high efficiency required for the acceleration of protons must be considered as a difficulty of the stellar flare origin of L-elements (the autogenic-spallative theory). This will be discussed again in Section V.

The destruction of Be by surface proton reactions, postulated to explain the low Be abundance and the high Li/Be ratio in F stars (section IV-7) rises no energetic problem. The destruction is limited to a very thin shell of matter (about $10^{-8}$ of the mass). Less than 10 eV per nucleon of the star are needed to destroy the Be of an F0–F2 star. Moreover this process would take place during the M.S. phase when nuclear energy sources are available and the allowed periods of time are considerable.

The rate of formation of L-elements at later periods of the life of a given star will, of course, depend on the high-energy particle flux and on the depth of the convective zone. To illustrate the situation we consider a specific problem. Grevesse (private communication) has pointed out that the Be lines in the solar optical spectrum are somewhat asymmetrical and that this asymmetry could be explained by assuming the presence of $^{10}$Be isotopes in the photosphere, with an isotopic ratio $^{10}$Be/$^9$Be of 0.1 to 0.2 (or $^{10}$Be = 0.2 to 0.5 in the conventional units). This ratio is about the formation ratio in spallation reactions (figures II-14 and II-15).

Since the isotope $^{10}$Be is unstable with half-life of $2.7 \times 10^6$ years, the implication (quite a surprise...) would be that the solar L-elements visible in the photosphere have been manufactured in the last few million years or so...

The amount of energy $\eta_L$ required to form one $^{10}$Be atom is $\simeq 2 \times 10^8$ MeV, and the surface convective mass is $\simeq 10^{-3}$ M$_\odot$. The total energy expenditure is then $\simeq 4 \times 10^{45}$ erg or $\simeq 4 \times 10^{31}$ erg/sec ($10^{-3}$ L$_\odot$) in the last few million years. By comparison the amount of energy coming out of the sun in the form of MeV particles, averaged over a solar cycle, is at best $10^{-9}$ L$_\odot$...

If the $^{10}$Be identification is confirmed, one or the other of the following possibilities will have to be considered:

a) The surface of the sun is a much more intense source of high-energy particles flux than we believe; only a minute fraction manage to escape the confining regions. (This is probably incompatible with the observations on solar X-rays.)

b) The L-elements do not, in $10^6$ years, manage to diffuse in the convective zone, they are retained in the (radiative) photosphere for very long periods.

At the present time, none of these two possibilities appears very likely.*

---

* Since then, this assumption has been abandonned.

Before we leave the treatment of the energy problems let us recall for completeness that a renewal of spallogeneration in the Red-Giant phase has been suggested (Feast) from the observations (figure IV-2). The evaluation of the energy requirements could be made within the frame of a stellar model specifying the time evolution of the depth and temperature of the surface convective zone. This way the "particle luminosity" characteristic of this period could be established. From the observation on the C stars (figure IV-2) it could be quite high.

# CHAPTER V

# Theories of the Origin of the Light Elements

IN THIS CHAPTER a summary and discussion of the various theories concerning the origin of the L-elements will be given. Two distinct problems arise in this context: a) what is the source of these light elements in young stars (the "initial" abundances of table IV-1), b) what is the source of the regeneration of L-elements at later phases of stellar evolution (if indeed such a regeneration does take place).

The table V-1 illustrates schematically a classification of these theories.

The "galactogenic" theories all assume that the initial L-elements are inherited, by the young star, from the interstellar gas (just as any other element). The "autogenic" theories assume that each star makes its own stock of these atoms; or more exactly that the contribution of the interstellar gas is much smaller than the "individual" contributions by the activity of the star under observation.

The one crucial observation to decide between these two classes of theories should come, of course, from a study of the interstellar gas itself. As seen in figures IV-1 and IV-16 we have, for the moment, only upper limits on the interstellar abundances of the L-elements. The upper limits on Li and Be are just comparable with the "initial" stellar values given in table IV-1 while the limit on D is about three times less than the solar system value (but the uncertainty is large). Clearly a definite decision should await more accurate studies.

**Table V-1**  A classification of the theories of L-elements generation in stars according to their characteristics. The thermonuclear theories involve very low energy particles (keV or so) while the spallative theories involve particles of tens or hundreds of MeV. In the galactogenic theories the L-elements are inherited from the interstellar gas, while in the autogenic theories they are generated in the very star in which they are observed

|  | Galactogenic | Autogenic |
|---|---|---|
| Thermonuclear $(E \ll 1 \text{ MeV})$ | "Big Bang" "Small Bangs" | Thermonuclear reactions and convection in the star itself |
| Spallative $(E \gg 1 \text{ MeV})$ | Supernovae Galactic Cosmic-Rays | Autoirradiation by stellar "flares" |

Another important observation on this respect is the following: the photosphere of the star T Tauri itself is reported to have much more Li than the surrounding expanding shell (Greenstein, Carter and Hudson). However the observation rises some difficulties which have not yet been properly explained (Herbig). There, again, the situation is not clear.

The test between the thermonuclear and spallative classes of theories is best made in term of the abundance ratios (see section II).

## V-1  THE THERMONUCLEAR THEORIES

In table V-1 are listed three types of thermonuclear theories. The "Big Bang" origin assumes that most of the L-elements were made in the very first (and hot...) moments of the universal expansion. The amount of L-elements generated this way can be calculated if one agrees with the view that the universal $3°K$ radiation is a remnant of the original "bang". One free parameter left is the "deceleration" rate. In figure V-1 the result of a calculation is illustrated (Wagoner).

The "little bangs" theory is concerned with the nucleo-synthesis within hypothetical massive objects evolving very rapidly from high temperatures ($>10^{10}$ K). Calculations have recently been performed by Wagoner.

In the frame of the *autogenic-thermonuclear* theories one considers the hypothetical transport to the surface by convective motion of the products of hydrogen shell-burning.

The equilibrium abundance of $^7Be$ in such a shell may, under special circumstances, be large enough to furnish (after beta-decay) an appreciable contribution of $^7Li$ at the surface.

This mechanism would, of course, contribute nothing to the "initial" abundances (it occurs far too late in stellar evolution) but could have something to do with the Red-Giants or Carbon stars L-elements regeneration (Cameron). No calculations are available.

The very large ratios of L-elements predicted by all thermonuclear theories makes them all highly inappropriate to explain the "initial" L-elements abundances. In particular the $^7Li/^6Li$ and the $^{11}B/^{10}B$ are always vastly different from the meteoritic ratios.

Partial effects may nevertheless exist. For instance "big bang" and "small bangs" contribution to D, $^3He$ (and perhaps $^7Li$) could be important (compare table IV-1 and figure V-1) (Mitler).

However the high frailty of the light elements makes it hard to believe that

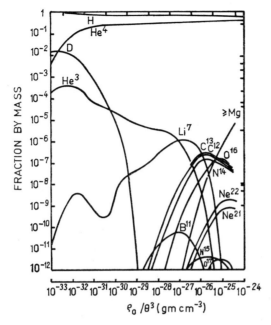

**Figure V-1** Element production in a universal fireball or a massive object expanding from $T > 10^9\,°K$. The abscissa represent the present baryon density ($\varrho_0$) divided by the present photon temperature ($\theta = T_0/3K$) where the $3K$ has been adopted from recent measurements at radio frequencies of a presumed universal background radiation (from Wagoner et al)

they could have survived from the birth of the Universe. More specifically, since those isotopes are readily destroyed in stellar interiors, the ratio of survival must be related to the fraction of the galactic gas which has never been trapped in a star. The determination of this ratio can only be made through a quantitative model of galactic evolution. Such models have been made by several authors, all of them involving rather large uncertainties.

A recent calculation made at Liège (P. Ledoux, private communication) suggests that a large fraction of the galactic mass has been, at one time or another, trapped in a star. This result is, of course, somewhat model dependent but, nevertheless, suggests that the "big bang" contribution of L-element to the present interstellar gas may be very small.*

---

* The opposite result has been recently obtained in a (somewhat different) model (Truran, private communication.)

## V-2   THE SPALLATIVE GALACTOGENIC THEORIES

The L-elements could be generated by spallation reactions during supernovae (SN) explosions (Cameron). The success of the theory of the supernovae origin of galactic cosmic-rays suggests strongly the presence of high energy particles (protons and nuclei) in the exploding mass or in the rotating atmosphere of the neutron star (pulsar) left over after the explosion (Gold, Peters). These particles could, indeed, induce the necessary reactions on the gas of the expanding shell. However, in absence of reliable SN models no calculation can be made at this point.

The elements could also be generated by the galactic cosmic-rays. This origin is clearly not independent of the previous one if the supernovae are

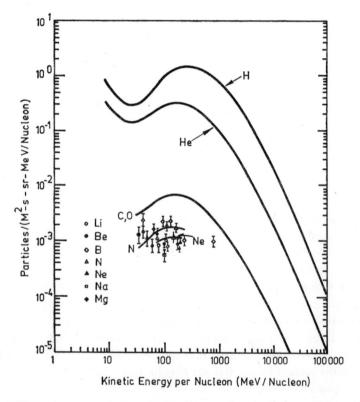

**Figure V-2**   Primary differential energy per nucleon spectra of cosmic-ray particles observed near solar minimum activity (from Simpson et al.)

responsible for the galactic cosmic-rays. We would have here two contributions from the same process: a "near" contribution – the spallation reactions on the expanding shell – and a "far" contribution – the spallation reactions generated by the galactic cosmic-rays on the interstellar gas. This last contribution is more amenable to calculations.

In figure V-2 recent measurements of the energy spectra of cosmic-ray particles near solar minimum activity are plotted. The "far" contribution is itself the sum of two different mechanisms: (a) the collisions of fast particles (mostly protons) with interstellar CNO, (b) the deceleration of L-elements formed by fast CNO collisions on the interstellar gas.

### V-2-1 The effect of galactic cosmic-ray protons*

The time-integrated proton flux required to generate the "initial" stellar abundance of L-elements is (see II-2-1)

$$\int \phi \, dt = \frac{n(L)}{\Sigma n \, (m) \, \sigma \, (m \rightarrow L)} \simeq 2 \times 10^{19} \text{ ptles cm}^{-2} \text{ sec}^{-1} \qquad \text{(V-1)}$$

From figure V-2 the present flux of protons capable of spallation i $\simeq 3.6$ ptles cm$^{-2}$ sec$^{-1}$. From radiochemical studies of unstable isotopes in meteorites it seems that this intensity did not vary much in the last billion years or so. If the flux remained the same throughout the life of the galaxy we get $10^{18}$ ptles cm$^{-2}$ sec$^{-1}$, or about 0.05 of the required value.

We must first take into consideration the fact that the flux in figure V-2 is the local flux near solar minimum and that the existence of a residual solar modulation is still present even at minimum (in other words, we are always "shielded" by the electromagnetic activity of the sun).

The calculation of the galactic cosmic-ray flux outside of the solar system from the known properties of the solar modulation is still a difficult and adventurous task. Numerous estimates exist of the ratio of fast protons ($E > 20$ MeV) in space and at the earth; they range from about three to hundreds... However it is fair to say that the more substantiated estimates (Webber) are closer to the lower limit.

The galactic flux could of course have been appreciably larger in the early days of the galaxy. But this rise of the flux was (at least partly) compensated by the lower value of CNO in the interstellar gas. And the survival problem of the L-elements must still be considered.

---

* This section stems from a remark by B. Peters.

### V-2-2  The deceleration of the L-elements

From figure V-2 the total flux of L-elements $\phi_L$ in the vicinity of the earth is $\simeq 4 \times 10^{-4}\,\mathrm{cm}^{-2}\,\mathrm{sec}^{-1}$ (integrated over the sphere). This represents a density of fast L atoms $n_L^f = \phi_L/V_L \simeq 3 \times 10^{-14}\,\mathrm{cm}^{-3}$.

A lithium atom of energy $E_0 \simeq 100\,\mathrm{MeV/N}$ will be brought to rest by ionization losses after crossing about one g cm$^{-2}$ of interstellar matter. (During this passage, hardly more than 20% of these atoms will have been destroyed by nuclear collisions.)

The deceleration time for this atom is approximately given by (if we use eq. III-8)

$$T_L \simeq \frac{4}{3}\,\frac{R_L(E_0)}{n(H)\,v_L(E_0)} = \frac{2 \times 10^6 y}{n(H)}, \quad \text{for } E_0 \simeq 100\,\mathrm{MeV/N} \quad (\text{V-2})$$

where $n(H)$ is the average density met by the incoming L atom (in cm$^{-3}$).

We can now compute the rate of injection of interstellar (slow) L atoms ($n_L^s$) in the galactic gas:

$$\frac{dn_L^s}{dt} = \frac{n_L^f}{T_L} = \frac{3}{4}\,\frac{\phi_L n\,(H)}{R_L} \simeq 5 \times 10^{-28}\,n(H) \quad (\text{V-3})$$

Assuming that the galactic cosmic-rays are confined to the galaxy ($n(H) \simeq 1$) and integrating over the life of the galaxy we get $n(Li)/n(H) \simeq 10^{-10}$. This is only a factor ten away from the initial stellar values. This estimate is subject to the same uncertainties as the estimate of the spallation of interstellar gas (V-2); uncertainties on the solar modulation, and on the time variation of the properties of the cosmic rays. At any rate this discussion shows that the SN theory of the L-elements (including both the "near" and the "far" contribution) remains a possibility.

### V-3  THE SPALLATIVE AUTOGENIC THEORY

Throughout this work the (generally accepted) view has been taken, that the "initial" stellar L-elements were generated by stellar flares in the very star in which they are seen.

This view is supported by a number of moderately good reasons, none of them being mandatory (or objection free): the signs of intense electromagnetic activity in T Tauri stars T Tauri itself—the upper limit of D/H in interstellar matter etc...

This theory, on the other hand, meets with an important difficulty: the energetics (section IV-9). The stellar flare origin of the "initial" stellar L-elements requires either that the flares be very hot (mean electron energies of 100 keV or so) or that most of the energy output of a T Tauri star be in the form of energetic protons (or both).

Also, the similarity between the solar and meteoritic $n(Li)/n(Si)$ is not automatically explained in the autogenic view. It appears as a requirement that any model of the solar system must meet.

# CHAPTER VI

# Summary

THE STELLAR OBSERVATIONS of lithium and beryllium can be accounted for by assuming that:

Early in the T Tauri period all stars are provided with an "initial stellar" abundance of L-elements, as given in table IV-1. These L-elements come from proton-induced spallation reactions on a cosmic gas (the most important targets are CNO), as witnessed by the isotopic ratios (compare with table II-2).

The lack of correlation between the distribution of Li as a function of spectral class (figure IV-1) for very young clusters and the fractional stellar mass in the surface convective zone as a function of spectral class (figure IV-8) show that the L-elements were given to the star when it was still largely convective (or before). We do not know yet if the initial L-elements were made by the star itself or if they were inherited from the galactic gas (in which case a supernovae-galactic cosmic-ray origin would be possible). A study of the interstellar gas should provide the answer.

After this initial contribution, very little L-addition takes place at least until the Giant-phase. The L-elements are depleted both by nuclear reactions (with protons) or by dilution with matter in which they have been previously destroyed (in the Giant phase).

Lithium isotopes are more vulnerable to keV protons than beryllium, but above $\sim 100$ keV the reverse is true.

The lithium isotopes are gradually destroyed in hot surface convective zone (most probable energies of a few keV). Pre-Main-Sequence depletion seems

---

*Note added in proof:*

As mentioned in the beginning the galactic cosmic-ray origin now appears to be the most promising one. The point is developed in "Galactic cosmic ray origin of Li, Be, B in stars", Reeves, H., Fowler, W. A., and Hoyle F., *Nature* **226,** 727, (1970), and its implication on various problems of galactic physics are discussed in "Spallation limits on instellar fluxes of low energy cosmic rays and nuclear gamma rays", (Fowler, W. A., Reeves, H. and Silk, J. Orange Aid preprint No. 209, appearing in *Astrophysical Journal* Oct. 70).

to be important for star later than G0. This may be related to the stellar dynamo effect of a rotating convective zone.

There are indications that beryllium is depleted by surface (100 keV to MeV) protons in stars with shallow convective zone during the M–S period.

During the Giant-phase the L-element abundances are first decreased by the onset of deep convective movements (dilution) and possibly later re-increased by a period of surface activity.

# Appendix

**Table A-1** (from Meyer) Cross section for the reaction $p + d \rightarrow p + p + n$

| $E_p$ (MeV) | $\sigma$ (mb) |
|:---:|:---:|
| 3.3 | 0 |
| 3.8 | 1.4 |
| 4.0 | 4.5 |
| 4.5 | 13 |
| 5.0 | 25 |
| 5.5 | 37 |
| 6.0 | 55 |
| 6.5 | 70 |
| 7.0 | 80 |
| 8.0 | 100 |
| 9.0 | 120 |
| 10.0 | 140 |
| 12.0 | 170 |
| 16.0 | 190 |
| 20.0 | 200 |
| 30.0 | 170 |
| 40.0 | 150 |
| 50.0 | 120 |
| 80.0 | 80 |
| 100 | 70 |
| 200 | 50 |
| 300 | 45 |

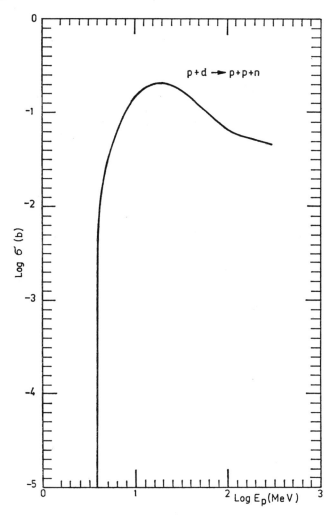

**Figure A–1.** (from Meyer)

| | $M_V$ | $T_e$ (°K) (1000°K) | B-V | Bolometric corrction | $M_{Bol}$ | Log $L/L_\odot$ | Log $M/M_\odot$ |
|---|---|---|---|---|---|---|---|
| 09 | −4.8 | | | | | | |
| B0 | −4.1 | 27 | −0.30 | −3.17 | −7.27 | 4.82 | |
| B1 | −3.5 | 23 | −0.26 | −2.50 | −6.00 | 4.3 | |
| B2 | −2.5 | 20 | −0.24 | −2.23 | −4.73 | 3.8 | |
| B3 | −1.7 | 18 | −0.20 | −1.71 | −3.41 | 3.3 | |
| B4 | | | | | | | 0.7 |
| B5 | −1.1 | 16 | −0.16 | −1.39 | −2.49 | 2.9 | 0.66 |
| B6 | | | −0.14 | −1.21 | | | |
| B6.5 | | 14 | | | | | |
| B7 | −0.6 | | −0.12 | −1.04 | −1.64 | 2.6 | 0.56 |
| B8 | −0.2 | 12.5 | −0.09 | −0.85 | −1.05 | 2.3 | 0.50 |
| B9 | +0.2 | 11.2 | −0.06 | −0.66 | −0.46 | 2.1 | 0.47 |
| A0 | +0.6 | 10.4 | 0.00 | −0.40 | +0.20 | 1.85 | 0.44 |
| A1 | 1.2 | 9.7 | 0.03 | −0.32 | +0.88 | 1.55 | 0.37 |
| A2 | | 9.1 | 0.06 | −0.25 | | | |
| A3 | 1.7 | 8.5 | 0.09 | −0.20 | 1.50 | 1.30 | 0.31 |
| A4 | | | | | | | |
| A5 | 2.1 | 8.2 | 0.15 | −0.15 | +1.95 | 1.15 | 0.26 |
| A6 | | | | | | | |
| A7 | 2.4 | 7.6 | 0.20 | −0.12 | +2.28 | 1.0 | 0.23 |
| A8 | | | | | | | |
| A9 | | | | | | | |
| F0 | 2.6 | 7.2 | 0.30 | −0.08 | 2.52 | 0.9 | 0.21 |
| F2 | 3.0 | 6.9 | 0.38 | −0.06 | 2.94 | 0.73 | 0.17 |
| F5 | 3.4 | 6.7 | 0.45 | −0.04 | 3.36 | 0.56 | 0.12 |
| F6 | 3.7 | 6.5 | 0.47 | −0.04 | 3.64 | 0.45 | 0.10 |
| F7 | | | 0.50 | −0.04 | | | |
| F8 | 4.0 | 6.2 | 0.53 | −0.05 | 3.95 | 0.33 | 0.06 |
| G0 | 4.4 | 6.0 | 0.60 | −0.06 | 4.34 | 0.17 | 0.03 |
| G2 | 4.7 | 5.74 | 0.64 | −0.07 | 4.63 | 0 | 0 |
| G5 | 5.2 | 5.52 | 0.68 | −0.10 | 5.10 | −0.13 | −0.05 |
| G8 | 5.6 | 5.32 | 0.72 | −0.15 | 5.45 | −0.27 | −0.09 |
| K0 | 5.9 | 5.12 | 0.81 | −0.19 | 5.71 | −0.38 | −0.11 |
| K1 | | 4.92 | | | | | |
| K2 | 6.3 | 4.76 | 0.92 | −0.25 | 6.05 | −0.51 | −0.14 |
| K3 | 6.9 | 4.60 | 0.98 | −0.35 | 6.55 | −0.71 | −0.20 |
| K5 | 8.0 | 4.35 | 1.18 | −0.71 | 7.29 | −1.0 | −0.29 |
| K6 | | 4.00 | | | | | |
| K7 | | | 1.38 | −1.02 | | | |
| M0 | 9.2 | 3.75 | | | | | |
| M1 | 9.7 | 3.60 | 1.48 | −1.70 | 8.0 | −1.2 | −0.39 |
| M2 | 10.1 | 3.55 | | −2.03 | 8.1 | −1.3 | −0.42 |
| M3 | 10.6 | 3.10 | 1.49 | −2.35 | 8.3 | −1.4 | −0.45 |
| M4 | 11.3 | | | −2.7 | 8.6 | −1.5 | −0.49 |
| M5 | 12.3 | | 1.69 | −3.1 | 9.2 | −1.8 | −0.58 |
| M6 | 13.4 | | | | | | |

# References

AUDOUZE, J., *Thesis* (in preparation).

AUDOUZE, J., and REEVES, H., "Destruction cross-sections of $^6$Li and $^7$Li by low energy protons", to be published in *Astrophys. J.*

AUDOUZE, J., EPHERRE, M., and REEVES, H., "Spallation of light nuclei", *Nucl. Phys.* **A97**, 144 (1967).

AUDOUZE, J., GRADSZTAJN, E., and REEVES, H., "The formation of lithium in young stars", *Les Congrès et Colloques de l'Université de Liège*, **41**, 299 (1967).

BAKER, N., *The Depth of the Outer-convection Zone in Main-Sequence Stars*, Institute for Space Studies, New York.

BASHKIN, S., and PEASLEE, D.C., "Production of the rare light elements". *Astrophys. J.* **134**, 981 (1961).

BERNAS, R., EPHERRE, M., GRADSZTAJN, E., KLAPISCH, R., and YIOU, F., *Phys. Letters* **15**, 147 (1965).

BERNAS, R., GRADSZTAJN, E., REEVES, H., and SCHATZMAN, "On the nucleosynthesis of lithium, beryllium and boron". *Annals of Physics* **44**, 426 (1967).

BETHE, H.A., *Ann. Physik* **5**, 325 (1930).

BETHE, H.A., and ASHKIN, J., in *Experimental Nuclear Physics*, (E. Segré Ed.), p. 166, Wiley, New York (1952).

BODENHEIMER, P., "Studies in stellar evolution II, lithium depletion during the pre-main sequence contraction". *Ap. J.* **142**, 451 (1965).

BODENHEIMER, P., "Depletion of deuterium and beryllium during pre-main sequence evolution", *Astrophys. J.* **144**, 103 (1966).

BÖHM, R.H., "The mixing of matter in the layer below the outer solar convection zone", *Astrophys. J.* **138**, 297 (1963).

BONSACK, W.K., and GREENSTEIN, J.L., "The abundance of Li in T Tauri stars and related objects", *Astrophys. J.* **131**, 83 (1960).

BONSACK, W.K., *Ap. J.* **133**, 340 (1961).

BRANDT, J.C., *Astrophys. J.* **144**, 1221 (1966).

BRETZ, M.C., *I.A.U. Symposium.* **26**, 304 (1964).

BURBIDGE, E.M., BURBIDGE, G.R., FOWLER, W.A., and HOYLE, F., "The synthesis of the elements in stars", *Rev. Mod. Phys.* **29**, 547 (1957).

CAMERON, A.G.W., *Astrophys. J.* **212**, 144 (1955).

CONTI, P.S., "Lithium destruction and rotational braking", Contribution from the Lick observatory n° 252.

CONTI, P.S., and WALLERSTEIN, G., "A search for lithium and beryllium in Ib Supergiants and Cepheides", Contributions from the Lick observatory n° 272.

CONTI, P.S., "Stellar Li/Be and $^6$Li/$^7$Li ratios", Contribution from the Lick Observatory n° 293.

CONTI, P.S., and WALLERSTEIN, G., "Lithium and beryllium in Stars", *Ann. Rev. of Astronomy*, vol. 7 (1970).

DAVIDS, C.M., LAUNER, H., and AUSTIN, S.M., "Production cross sections of light elements from proton spallation of carbon". Preprints.

DAVIDS, C.M., LAUNER, H., and AUSTIN, S.M., *Phys. Rev. Let.*, **22**, 1388 (1969).

DANZIGER, T.J., and CONTI, P.S., *Astrophys. J.* **146**, 383 (1966).

DANZIGER, T.J., *Astrophys. J.* **150**, 733 (1967).

DEMARQUE, P., and ROEDER, R.C., *Astrophys. J.* **147**, 1188 (1967).

EBERHARDT, P., GEISS, J., and GRÖGLER, N., *J. Geophys. Res.* **72**, 331 (1967).

EPHERRE, M., and REEVES, H., "Speculations on the formation of light elements in stars", *Astrophysical Letters*, to be published.

ERICSON, T., *Phil. Mag. Suppl.*, **9**, 425 (1960).

EZER, D., and CAMERON, "A study of solar evolution", *Canadian Journal of Physics* **43**, 1497 (1965).

FAN, C.Y., COMSTOCK, G.M., and SIMPSON, J.A., *Proc. 10th Int. Cosmic Ray Conf. Calgary* (1968), *Can. J. Phys.* **46**, S 548 (1968).

FEAST, M.W., *Highlights of Astronomy*, editor L.Perek 233 (1966). Reidel Pub. Co., Dordrecht, Holland.

FICHTEL, C.E., and MACDONALD, F.B., *Ann. Rev. of Astronomy and Astrophysics* **5**, 351 (1967).

FOWLER, W.A., BURBIDGE, G.R., and BURBIDGE, E.M., "Nuclear reactions and element synthesis in the surfaces of the stars", *Astrophys. J. Suppl.* **2**, 167 (1955).

FOWLER, W.A., GREENSTEIN, J.L., and HOYLE, F., "Nucleosynthesis during the early history of the solar system", *Geophys. J. R.A.S.* **6**, 148 (1962).

FOWLER, W.A., CAUGHLAN, G.R., and ZIMMERMAN, B.A., *Ann. Rev. of Astron. and Astrophys.* **5**, 525 (1967).

FRIEDLANDER, KENNEDY and MILLER, *Nuclear Radiochemistry*, John Wiley and Son Inc., New York.

FRIEDMAN and WEISSKOPF, "The compound nucleus", in *Niels Bohr and The Development of Physics*.

GOLD, T., *Dallas Conference* (1968).

GOLDREICH, P., and SCHUBERT, G., "Differential rotation in stars", *Astrophys. J.* **150**, 571 (1967).

GRADSZTAJN, E., These Orsay, April 1965. *Ann. Phys.* (Paris), **10**, 791 (1965).

GREENSTEIN, J.L., and RICHARDSON, R.S., "Lithium and the internal circulation of the sun", *Astrophys. J.* **113**, 356 (1951).

GREENSTEIN, J.L., *Astrophys. J.* **113**, 531 (1951).

GREVESSE, N., "Solar abundances of lithium beryllium and boron", *Solar Physics* **5**, 159 (1968).

HAYAKAWA, S., *Prog. Theoret. Phys.* **13**, 464 (1955).

HAYAKAWA, S., "Origin of light elements in solar system". Preprint (1968).

HAYASHI, C., HOSHI, R., and SUGIMOTO, D., *Progr. Theor. Phys. Suppl.* n° 22 (1962).

HAYASHI, C., "Evolution of protostars", *Ann. Rev. of Astronomy and Astrophysics* **4**, 171 (1966).

HERBIG, G.H., *Astrophys.* **141**, 588 (1965).

HERBIG, G.H., and WOOLF, R.J., *Ann. Astrophys.* **29**, 593 (1966).

References 85

HERBIG, G.H., "Lithium in main-sequence stars", p. 411 of *Stellar Evolution* editors R.F. Stein and A.G.W.Cameron, Plenum Press (1966).

HERBIG, G.H., "On the interpretation of FU Orionis" in *Vistas in Astronomy* editor A.Beer **8**, 109 (1966).

HOWARD, L.N., MOORE, D.W., and SPIEGEL, E.A., *Nature* **214**, 1297 (1967).

HUDSON, R.D., and CARTER, V.L., *Astrophys. J.* **143**, 263 (1966).

IBEN, I., *Astrophys. J.* **142**, 1447 (1965).

IBEN, I. Jr., "Stellar evolution within and off the main sequence", *Ann. Rev. of Astronomy and Astrophysics.* Vol. 5, p. 571 (1967).

JUNG, M., *Thèse de 3ème cycle*, Université de Strasbourg, France (1968).

KINMAN, T.D., *Monthly Notices* **116**, 77 (1956).

KRAFT, R.P., and WILSON, O.C., *Astrophys. J.* **141**, 828 (1965).

KRAFT, R.P., *Astrophys. J.* **150**, 551 (1967).

KRANKOWSKY, D., and MÜLLER, O., "Isotopic composition and abundance of lithium in meteoritic matter", *Geochim. and Cosmochim. Acta* **31**, 1833 (1967).

KUHI, L.V., "Mass loss from T Tauri stars", *Astrophys. J.* **140**, 1409 (1964).

LAL, D., and RAJAN, R.S., "Observations relating to space irradiation of individual crystals of gas rich meteorites", *Nature*, to be published (1969).

LAMBERT, D.L., and WARNER, B., *Monthly Notices Roy. Astron. Soc.* **138**, 181 (1968).

LEFORT, M., *La Chimie Nucléaire*, Dunod (1967).

McKELLER, A., *Pub. Astron. Soc. Poc.* **52**, 407 (1940).

MERCHANT, A.E., "Beryllium in F and G type dwarfs", *Astrophys. J.* **143**, 336 (1966).

MERCHANT, A.E., *Astrophys. J.* **147**, 587 (1967).

METROPOLIS, N., BIVINS, R., STORM, M., TURKEVICH, A., MILLER, J.M., and FRIEDLANDER, G., *Monte Carlo calculations on intranuclear cascades*, *Phys. Rev.* **110**, 185 (1958).

MEYER, J.P., "Création de deutérons et d'³He au cours de la propagation interstellaire du rayonnement cosmique", *Thesis Saclay* (1969).

MILLER, J., and HUDIS, A., *Ann. Rev. Nucl. Sci.* **9**, 159 (1959). "Studies in penetration of charged particles in matter" Publication 1133, National Academy of Sciences, Washington, D.C. (1964).

OKAMOTO, I., "On the loss of angular momentum from the protosun and the formation of the solar system", *Publications of Astronomical Society of Japan*, to be published.

PARKER, E., *Astrophys. J.* **121**, 491 (1955).

PARKER, P.D., BAHCALL, J.N., and FOWLER, W.A., "Termination of the proton-proton chain in stellar interiors", *Ap. J.* **139**, 602 (1966).

PEIMBERT, M., and WALLERSTEIN, G., "A search for deuterium in Stellar spectra" I and II, *Astrophys. J.* **142**, 3 (1965); **141**, 2 (1965).

PELLAS, P., POUPEAU, G., LORIN, J.C., REEVES, H., and AUDOUZE, J., "Primitive low-energy particle irradiation of meteoritic crystals", *Nature*, to be published (1969).

PEPIN, R.O., and SIGNER, P., "Primordial rare gases in meteorites", *Science* **149**, 253 (1965).

PETERS, B., to appear in *Physique Fondamentale et Astrophysique*. Nice (April 1969), *Journal de Physique* **C3**, (Nov. 1969), Paris.

POVEDA, A., "On the nature of the infrared excesses in T-Tauri like stars", unpublished.

POVEDA, A., "The H.R. diagrams of young clusters and the formation of planetary systems", Publication of the Observatory of Mexico.

QUIJANO-RICO, M., and WANKE, H., "Determination of boron lithium and chlorine in meteorites", *Proceedings of the conference Research on meteorites, Vienna* (1968).

REEVES, H., "A review of nuclear energy generation in stars and some aspects of nucleo-synthesis" 83, of "Stellar Evolution", editors R.F. Stein and A.G.W.Cameron, Plenum Press (1966).

REEVES, H., "High energy nucleosynthesis" lectures given at the Summer School on "High Energy Astrophysics" at Les Houches (1966) and at the C.N.E.S. (Tarbes, 1966).

REEVES, H., *Stellar Evolution and Nucleosynthesis* (Lectures given at l'Université de Bruxel-les, 1964, and at l'Institut d'Astrophysique de Paris, 1965, Gordon and Breach, New York (1968).

REEVES, H., and AUDOUZE, "The spectrum of accelerated particles in stars, from lithium and beryllium observations", *Astrophysical Letters* **1**, 197 (1968).

RUDSTAM, G., *Thesis*, Uppsala, Sweden (1956).

RUDSTAM, G., "Systematics of spallation yields", *Zeit. Natf.* **21a**, 1027 (1966).

RYTER, C., REEVES, H., GRADSZTAJN, E., and AUDOUZE, J., "The energetics of light nuclei formation in stellar atmospheres", to be published.

SCHATZMAN, E., "Role of magnetic activity during stellar formation", from *Stellar Evolution*, R. F. Stein and A. G.W. Cameron editors, Plenum Press, New York (1966).

SCHATZMAN, E., *Studies in Stellar Evolution*, City College of New York (1967).

VON SENGBUSCH, K., Sternentwicklung IX. "Die erste hydrostatische Kontraktionsphase für einen Stern von 1 M$_\odot$", *Zeitschrift für Astrophysik* **69**, 79 (1968).

SERBER, R., *Phys. Rev.* **72**, 1114 (1947).

SHEN, B.S.P., *High Energy Nuclear Reactions in Astrophysics*, Editor W.A.Benjamin, Inc. New York, Amsterdam.

SHIMA, M., and HONDA, M., *J. Geophys. Res.* **68**, 2849 (1963).

SILL, C.W., and WILLIS, C.P., *Geochim. Cosmochim. Acta* **26**, 1209 (1962).

SPITZER Jr., L., *Physics of the Fully Ionized Gases*, Interscience Publishers Inc., New York (1956).

TORRES PEIMBERT, S., WALLERSTEIN, G., and PHILLIPS, J.G., "Lithium in carbon stars", *Astrophys. J.* **140**, 1315 (1964).

VAN DEN HEUVEL, E.P.J., *A Study of Stellar rotation*, North Holland Publishing (1968).

WAGONER, R.V., FOWLER, W.A., and HOYLE, F., "On the synthesis of elements at very high temperatures", *Astrophys. J.* **148**, 3 (1967).

WAGONER, R.V., "Explosive nucleosynthesis", *Ap. J.* **151**, L103 (1968).

WALLERSTEIN, G., HERBIG, G.H., and CONTI, P.S., "Observations of the lithium content of main-sequence stars in the Hyades", *Astrophys. J.* **141**, 610 (1965).

WEBBER, E.J., and DAVIS, L. Jr., *Astrophys. J.* **148**, 217 (1967).

WEBBER, W.R., "On the relationships between recent measurements of cosmic ray electrons, non thermal radio emission from the galaxy, and the solar modulation of cosmic rays". *Aust. J. Phys.* **21**, 845 (1968).

WEINREB, S., *Nature* **195**, 367 (1962).

WEYMANN, R., and SEARS, R.L., "The depth of the convective envelope on the lower main-sequence and the depletion of lithium", *Astrophys. J.* **142**, 174 (1965).

WIEHR, E., STELLMACHER, G., and SCHROTER, E.H., "On lithium in sunspots", *Astrophysical Letters* **1**, 181 (1968).